Human Predators and Prey Mortality

Human Predators and Prey Mortality

EDITED BY
Mary C. Stiner

Westview Press
BOULDER • SAN FRANCISCO • OXFORD

GN
799
F6
H86
1991

Westview Special Studies in Archaeological Research

3 1254 01319 6403

This Westview softcover edition is printed on acid-free paper and bound in library-quality, coated covers that carry the highest rating of the National Association of State Textbook Administrators, in consultation with the Association of American Publishers and the Book Manufacturers' Institute.

All rights reserved. No part of this publication may be reproduced or transmitted in any form or by any means, electronic or mechanical, including photocopy, recording, or any information storage and retrieval system, without permission in writing from the publisher.

Copyright © 1991 by Westview Press, Inc.

Published in 1991 in the United States of America by Westview Press, Inc., 5500 Central Avenue, Boulder, Colorado 80301, and in the United Kingdom by Westview Press, 36 Lonsdale Road, Summertown, Oxford OX2 7EW

Library of Congress Cataloging-in-Publication Data
Human predators and prey mortality / edited by Mary C. Stiner.
  p. cm. — (Westview special studies in archaeological research)
  Includes bibliographical references.
  ISBN 0-8133-8365-X
  1. Man, Prehistoric—Food. 2. Hunting and gathering societies.
3. Animal remains (Archaeology). 4. Predation (Biology). I. Stiner, Mary C., 1955- . II. Series.
GN794.F6H86 1991
306.3—dc20   91-13394
                                                              CIP

Printed and bound in the United States of America

∞ The paper used in this publication meets the requirements of the American National Standard for Permanence of Paper for Printed Library Materials Z39.48-1984.

10 9 8 7 6 5 4 3 2 1

Contents

List of Tables vii
List of Figures xi
Acknowledgments xvii

1. Introduction: Actualistic and Archaeological
   Studies of Prey Mortality 1
   Mary C. Stiner

2. Hunting Strategies, Prey Behavior and Mortality Data 15
   George C. Frison

3. Taphonomy and Early Hominid Behavior: Problems in
   Distinguishing Cultural and Non-Cultural Agents 31
   John D. Speth

4. Examining and Refining the Quadratic Crown Height
   Method of Age Estimation 41
   Diane Gifford-Gonzalez

5. Procurement Technology and Prey Mortality Among
   Indigenous Neotropical Hunters 79
   Michael Alvard and Hillard Kaplan

6. Nonselective Small Game Hunting Strategies: An
   Ethnoarchaeological Study of Aka Pygmy Sites 105
   Jean Hudson

7. Prey Size and Age Models of Prehistoric Hominid
   Scavenging: Test Cases from the Serengeti 121
   Robert J. Blumenschine

8. An Interspecific Perspective on the Emergence
   of the Modern Human Predatory Niche 149
   Mary C. Stiner

9. Subsistence Change and Pinniped Hunting 187
   R. Lee Lyman

10. Thule Eskimo Subsistence and Bowhead Whale
    Procurement 201
    James M. Savelle and Allen P. McCartney

11. Seasonality Studies and Paleoindian
    Subsistence Strategies 217
    Lawrence C. Todd

References Cited 239
List of Contributors 277

Tables

4.1 Actual age distributions of bison in the Wheatland and Yellowstone dental samples 55

4.2 Two-sample t-test of molar breadth for Wheatland and Yellowstone samples 56

4.3a Two-sample t-test of M1 crown heights, by age class, for Wheatland and Yellowstone samples 57

4.3b Two-sample t-test of M2 crown heights, by age class, for Wheatland and Yellowstone samples 57

4.3c Two-sample t-test of M3 crown heights, by age class, for Wheatland and Yellowstone samples 58

4.4 Summary of probabilities of results, Kolomogorov-Smirnov tests comparing Wheatland and Yellowstone mortality profiles constructed from actual ages at death 72

5.1 Comparison of hunting efficiency between shotguns and bows 88

5.2a Shotgun pursuits of large terrestrial mammals and arboreal primates during hunts observed near Diamante 89

5.2b Bow pursuits of large terrestrial mammals and arboreal primates during hunts observed near Yomiwato 91

5.3 Mean and standard deviation of pursuit times for terrestrial mammals and arboreal primate prey 93

5.4 Age composition of harvested prey by species obtained by shotgun hunters and by bow-hunters compared with census data for free-ranging populations 95

5.5	Sex of selected prey species killed near Diamante and Yomiwato and sex ratios based on census data for free-ranging populations	97
6.1	List of species recovered archaeologically from two modern Aka camps	117
7.1	Size profiles for the Serengeti carcass sample	127
7.2	Definition of age groups based upon eruption and wear of mandibular teeth, and equated percentages of lifespan for the blue wildebeest and the white-bearded wildebeest	136
7.3	Age profiles for wildebeest in the Serengeti carcass sample	139
7.4a	Age profiles for wildebeest killed by lions in the Serengeti carcass sample and as documented by Schaller	144
7.4b	Correspondence between the age classifications used for the carcass sample and Schaller's kill sample	144
8.1	Background information for the Italian Paleolithic sites	152
8.2	Median percent values for the three ungulate age cohorts by case groups	168
8.3	Regression statistics for the relationship between H+H/L and tMNE/MNI in faunal assemblages collected by predators at shelters	172
8.4	Summary statistics for H+H/L ratio by predator type	174
8.5	Head to limb and prime- to old-aged adult ratios for medium ungulate remains in selected Italian Paleolithic shelter faunas	178
9.1a	Frequencies of seven mammalian taxa in three assemblages dating between 3300 and 2000 B.P.	192
9.1b	Frequencies of seven mammalian taxa in five assemblages dating between 2000 and 200 B.P.	193

Tables ix

9.2 Frequencies of harpoon parts and arrow heads associated with Oregon Coast faunal assemblages 195

11.1 Dental ages and estimated season of death at Paleoindian bison bonebeds 221

11.2 Relative frequencies of mature male metapodials from several Paleoindian bonebeds 227

Figures

1.1　Two basic age structure models exemplified using nine age cohorts 7

2.1　The way not to jump bison 17

2.2　Domestic cattle in bogs 20

2.3　Decision-making time while handling an unhappy bison cow 21

2.4　Mountain sheep under a net 24

2.5　African elephant family herd at a watering hole 26

4.1　Differences in quadratic-derived age estimations for molars in Neolithic cattle partial dental arcades 48

4.2　Mortality profiles for Neolithic cattle teeth from GvJm44, Lukenya Hill, Kenya 50

4.3　Mortality profiles for Neolithic cattle teeth from Narosura, Loita Plains, Kenya 52

4.4　Percent error of quadratic-derived age estimates for M1, Wheatland sample, grouped by actual ages 60

4.5　Percent error of quadratic-derived age estimates for M1, Yellowstone sample, grouped by actual ages 61

4.6　Percent error of quadratic-derived age estimates for M2, Wheatland sample, grouped by actual ages 62

4.7　Percent error of quadratic-derived age estimates for M2, Yellowstone sample, grouped by actual ages 63

4.8 Percent error of quadratic-derived age estimates for M3, Wheatland sample, grouped by actual ages 64

4.9 Percent error of quadratic-derived age estimates for M3, Yellowstone sample, grouped by actual ages 65

4.10 Curvilinear regression of Yellowstone M1 data compared to theoretical model 66

4.11 Mortality profiles for Wheatland sample: from actual ages and from quadratic-derived estimates for the three molars 70

4.12 Mortality profiles for Yellowstone sample: from actual ages and from quadratic-derived estimates for the three molars 71

4.13 Relation of basal enamel breadth to initial crown height, combined Wheatland and Yellowstone samples 75

5.1 Map of the Manu National Park area 82

6.1 Two hypothetical mortality profile models: catastrophic and selective prime-age hunting 108

6.2 Mortality profile for the archaeological assemblage of blue duikers net-hunted by Aka foragers 109

6.3 Age profile based on live census data for a wild population of blue duikers, also captured with nets 110

6.4 Comparison of the results of three alternative methods for constructing mortality profiles from Aka archaeological assemblages 114

6.5 Taxonomic composition of hunted game remains at two modern Aka camps, based on MNI values for recovered bone 116

6.6 Size classes of hunted game remains at two modern Aka camps, based on MNI values for recovered bone 119

Figures xiii

7.1 The relationship between live weight of mammalian herbivores and their persistence upon death as sources of scavengeable food in the Serengeti 126

7.2 Larger mammalian herbivore size distributions of the live community and carcass sample from the Serengeti 129

7.3 Size distributions of scavengeable carcasses encountered in different diurnal contexts in the Serengeti 131

7.4 Ideal catastrophic and attritional age profiles based on increments of life-span, and transformed into four stages 135

7.5 Age profiles of wildebeest carcasses encountered in different diurnal contexts in the Serengeti 140

7.6 Age distribution of wildebeest killed by lions 143

8.1 Occlusal wear stages for deciduous and permanent lower 4th premolars of sheep/goat, cattle and deer 154

8.2 Three idealized mortality profiles found in nature, superimposed on a stable living population model 158

8.3 Triangular diagrams defining the three age axes, and relating areas of the graph to general classes of mortality pattern 159

8.4 Triangular plot of ungulate mortality cases generated by modern cursorial and ambush carnivores 160

8.5 Triangular plot of mortality patterns in cervid and bovid remains in North American Holocene archaeofaunas and modern trophy hunting 163

8.6 Triangular plot of mortality patterns in cervid and bovid remains in Paleolithic Italian shelters 164

8.7 Triangular plot of median mortality values for modern predators grouped by strategy class or environment and culture 166

8.8   Triangular plot of median mortality values for various
      modern and Paleolithic predator groupings 167

8.9   Log-log regression of anatomical completeness against the
      proportion of head and horn parts to limbs for ungulate
      remains at shelters 170

8.10 Log-log regression of anatomical completeness against the
      proportion of head and horn parts to limbs collected by
      hyaenas and wolves and by Paleolithic hominids 173

8.11 Log-log regression of the proportion of head and horn parts
      to limbs against the proportion of prime-age to old
      individuals in Italian Paleolithic death assemblages 177

9.1   Modeled variation in the relative proportions of resources
      exploited by foragers and collectors 191

9.2   Observed variation in the proportions of seven mammalian
      taxa for three Oregon coast assemblages 194

10.1 Bowhead whale and skeleton 203

10.2 Thule Eskimo site distribution and archaeological
      bowhead whale bone abundance 208

10.3 Relative length frequencies of bowhead whales in live
      populations in the Beaufort Sea, in stranded Holocene
      populations, on Brodeur Peninsula and Baffin Island, and
      in Thule sites in all survey regions 211

10.4 Relative length frequencies of bowhead whales represented
      at Thule sites in core and peripheral summer ranges 213

11.1 Locations of selected Paleoindian bonebeds for which
      seasons of death have been established 220

11.2 Seasons of death based on eruption and wear patterns of
      mandibular dentitions from Paleoindian bonebeds 223

Figures xv

11.3 Seasons of death in Paleoindian faunas, based on the frequency of occurrence of 0.1 year increments 225

11.4 Differences in the percentages of mature male metacarpals and metatarsals relative to females 228

11.5 Model of relationships between changes in somatic growth potential and the length of time in which bison fat resources are available to human predators 233

11.6 The four compartments of a ruminant stomach 236

Acknowledgments

This group of papers was first presented in a symposium entitled "Applications of Mammalian Mortality Patterns for the Study of Prehistoric Human Foraging Ecology" at the 54th Annual Meeting of the Society for American Archaeology in Atlanta, Georgia, April 1989, and I am grateful for the society's support in providing a forum. I thank the contributors to the volume for their interest and diligence, as they are responsible for most of the positive qualities of this book. Much credit is also due to June-el Piper, whose technical assistance in preparing the camera-ready manuscript for publication has been absolutely crucial. I extend thanks also to Kellie Masterson and Ellen McCarthy of Westview Press for their help in editorial and technical matters and to Nancy Stone for creative input in naming the book — suffice it to say, the least colorful of the many choices serves as the final title.

This document was produced using WordStar Version 6.0 and Times Roman fonts on a PSJet+ Laser Printer. The following illustrations first appeared in other publications and are reprinted here with permission: Figures 2.3 and 2.4 (pages 21 and 24) originally appeared in George C. Frison in *The Evolution of Human Hunting* edited by M. H. Nitecki and D. V. Nitecki (New York: Plenum Press, 1987), Figs. 4 and 14; Figures 8.1 through 8.6 (pages 154, 158, 159, 160, 163, and 164) originally appeared in Mary C. Stiner *Journal of Anthropological Archaeology* 9, Figs. 2, 3, 6, 8, and 9; Figures 9.1 and 9.2 (pages 191 and 194) originally appeared in R. Lee Lyman, *Prehistory of the Oregon Coast* (Orlando, Fla.: Academic Press, 1991), Figs. 4.2 and 8.2.

M.C.S.

1

Introduction: Actualistic and Archaeological Studies of Prey Mortality

Mary C. Stiner

Age structures, or mortality patterns, in mammalian archaeofaunas reflect some very basic ecological relationships between humans and their prey. Mortality data have considerable potential for addressing questions about human land use, food search practices, the influence of technology on prey acquisition, labor organization, and the very nature of the human foraging niche itself. However, while mortality studies appear in many publications, their highly technical nature and limited scope often isolate them from the rest of archaeological research. This book attempts to bring a promising approach into the mainstream of behavioral issues in anthropology.

Mortality data represent only one dimension of human foraging practices, yet they interest many zooarchaeologists for a number of reasons. One is the very practical issue of preservation: mammal teeth are among the most persistent materials in archaeofaunas and they are the preferred bases for aging. Dentitions undergo progressive, measurable changes over the entire lifetime, first through growth and later through occlusal wear and the formation of cementum layers. Living bone also undergoes alterations as a function of individual age, but easily perceived changes in these tissues generally cease with skeletal maturity.[1]

A second reason for archaeologists' continuing interest in mortality data stems from the fact that ungulate and other large mammal remains are common in prehistoric faunas. Whether hunters or scavengers, the record of hominid predation on relatively large animals extends back roughly 2 million years. Relationships between humans and their prey obviously have changed greatly over this vast time range, but hominid foraging systems were always affected in some way by the constraints imposed by mammalian resources. Because certain predictable links exist between the age structures of prey death assemblages and general ecological principles of predator-prey dynamics, mortality patterns present an opportunity to "map" the diversity in modern human exploitation of mammals, as well as humans' changing place in animal communities in prehistory.

The volume centers on identifying and distinguishing human predatory behaviors from patterns of prey age selection. As a group, the studies place a strong emphasis on control research conducted in the present and application of uniformitarian assumptions about mammalian behavior for studying the past. They also integrate mortality data with other classes of information, offering detailed accounts of how age patterns in death assemblages can be influenced by human weaponry and cooperation, as well as by species-specific differences in prey behavior, physiology and distributions. The presentations are designed to aid archaeologists in both teaching and research contexts. They are also likely to interest anthropologists concerned with humans' impact on mammalian populations and human subsistence ecology more generally.

This introductory chapter discusses several issues relevant to how mortality studies are conducted and what they can reveal about human behavior. Foremost among these is the continuing need for parallel investment in actualistic, cross-cultural and inter-specific comparative studies. A second set of issues concerns what mortality data are and the basic theoretical models integral to their use. The final points to be discussed concern scales of observation, their potential relevance to specific behavioral phenomena, and the use of control populations as comparative baselines for studying human predatory habits.

RATIONALE AND ORGANIZATION OF THE VOLUME

In reviewing the analytical potential of mortality data several years ago, I was both surprised and frustrated by the lack of clear logical connections between pattern and cause. A great disparity currently lies between the technically sophisticated means for constructing mortality profiles and the knowledge available to interpret these patterns. Perhaps my concern would not have been so acute had I not been working with

Introduction

Middle Paleolithic faunas. In any archaeological situation involving premodern hominids, it is quite likely that there are no modern analogs for the foraging habits under study. In lieu of direct analogs, more general ecological principles for interpreting the mortality data are needed. It also is clear that closing this gap will be a major undertaking, requiring the cooperation of many researchers working in many contexts, and that control studies must extend beyond an exclusive focus on humans.

With these issues in mind, I invited a diverse group of anthropologists to participate in identifying some of the major problems in mortality studies with regard to human predatory behavior and to strengthen this branch of research with actualistic and comparative data. The papers illustrate some novel shifts in view, sometimes diverging from or even confronting interpretive conventions set forth previously. The contributors nevertheless recognize the historical importance of earlier research on the topic of prey mortality and human predation as the foundations for developing their own: the works of Ducos, Jarman, Klein, Kurtén, Payne, Reher, Severinghaus, B. Smith, Spiess, Wilkinson, and Voorhies have been especially important in this regard.

The Scope of Inquiry

It must be acknowledged that mortality studies have wider applications than those represented here; the contributions in this volume focus on exploitation of wild prey by hunter-gatherers and small-scale horticulturalists. Issues surrounding animal domestication strategies are categorically avoided because they usually involved deliberate redirection of energy by human consumers back into prey populations. In contrast, simpler economies based on hunting and scavenging simply exploit the entropy in ecosystems without significant reinvestment, a trophic characteristic shared with nonhuman carnivores.

The volume's scope nonetheless is inspired by Jarman and Wilkinson's (1972) insightful essay on how best to investigate problems of animal domestication from the perspective of mortality data. These authors recognized that alterations in prey skeletal morphology could only provide signatures of the later phases of certain kinds of domestication. Mortality patterns were seen as a potentially powerful alternative for investigating earlier stages of the same processes, but that their analytical utility was contingent upon knowledge of causality and natural variation in population composition outside of the human sphere. Jarman and Wilkinson were among the first to outline just how and where archaeologists must push the boundaries of knowledge about mortality phenomena in order to realize their full potential for behavioral research. Their recommendations pointed largely to the areas of prey behavior and physiology and to predator-prey dynamics. Together, the papers assembled here respond to this challenge.

Parallel Investment

Archaeology is, by definition, about the past, yet interpretations of archaeofaunas are based on our perceptions of how things work in the present. Both actualistic and archaeological data are mandatory for setting observed mortality patterns into a rigorous interpretive framework. Actualistic data delimit the relevant domains of causality that can shape mortality patterns in death assemblages, thereby allowing one to know the degree to which any pattern could possibly conform with specific kinds of predatory behaviors. Each author in this volume conducts some background work in the modern world, seeking reliable links between mortality phenomena, methods of aging animals, and behavioral and developmental characteristics of prey, as well as exploring associations with independent data sets such as technology.

These approaches are not intended to provide procedural recipes. They illustrate a rich array of methodological options and, as a group, provide many suggestions about how to solve problems in human behavioral ecology for which there are no ready and convenient solutions. The articles also document some simple yet very important facts about humans as predators. Not least among these is the observation that humans living in traditional societies are subject to many of the same basic constraints that affect nonhuman predators, and thus human responses often can be modeled using more general ecological principles. For example, hunters respond to prey abundance and distributions to a large degree. Prey behavior, body size and tissue composition also exercise marked effects on human strategies. To the extent that it can be addressed here, hunters' choices and processing decisions appear responsive to fat content in prey, depending on the overall availability of this key resource in the environment. Finally, cooperation and certain types of technology can enhance the range of options hunters have in choosing prey items.

The first three papers in this book are methodological essays. One concerns relationships between prey behavior, human procurement strategies and mortality patterns in archaeofaunas (Frison). The next considers problems of identifying human responses to seasonal changes in prey nutritional state and prey age as reflected by procurement and processing decisions (Speth). The third paper explores natural variation in tooth development and wear that potentially distorts our perceptions of the age structures of death populations (Gifford-Gonzalez).

Topics then shift to actualistic studies of modern human foragers, their potential relevance for modeling prehistoric human predatory habits (Alvard and Kaplan) and the impact of socioterritorial adaptations of forest species on prey age "selection" by humans (Hudson). These two studies explore the impact of technology, use of dogs, and prey body size and gender on procurement in tropical forest environments. Because they consider prey species outside the usual array considered in mortality research (including primates, tapirs, peccaries and small antelopes), they also aid in establishing potential variation in contemporary human foraging practices.

The next two chapters explore the strategic implications of variation in mortality phenomena for changes in Pleistocene hominid foraging niche. This is accomplished through actualistic research on scavenging opportunities in some East African habitats (Blumenschine) and through interspecific comparisons of prey age selection and food transport by human and nonhuman predators across several regions of the New and Old Worlds (Stiner). Such studies aid in isolating some of the ecological (and evolutionary) mechanisms relevant to understanding changes in prehistoric hominid foraging niche.

The final trio of papers concerns specific applications of actualistic, ethnographic and ethological data to research on archaeofaunas. These studies consider three very different mammalian orders (pinnipedia, cetacea and artiodactyla) and examine the potential interactive effects of prey ranging behavior, life history characteristics, seasonal cycles in nutritional state, and prehistoric human procurement strategies. Seals (Lyman) and whales (Savelle and McCartney) were important in many high latitude human economies, and changes in the relative importance of marine mammals and the modes of exploitation are traced in part through mortality patterns. Both of these studies document the special "window-like" effects that often characterize marine exploitation by a terrestrial predator. In the case of seals, procurement on land and prey choice are contingent upon the reproductive patterns of each species. Body size is a major determinant of prey choice in hunting whales, whereas a very different set of factors conditions the age structure of scavengeable individuals. In the last paper, seasonal mortality in bison is used to refine current hypotheses about Paleoindian over-wintering strategies in a Late Pleistocene/Early Holocene grassland environment (Todd). While the only paper with seasonality as its main theme, it sets a compelling example for modeling human responses to the amplitude and intensity of seasonal changes in resource availability.

WHAT ARE MORTALITY PATTERNS AND WHERE DO THEY COME FROM?

The papers in this volume assume a basic knowledge of what mortality patterns are and how they vary. In order to avoid repetition, this section summarizes the principles, variables and models fundamental to age structure analyses. Two mortality models that appear throughout the book are based on simple, well-documented age structure patterns (e.g., Caughley 1966, 1977). They have nearly universal applicability for mammals because they are the direct products of recruitment (live births), development rates and survivorship. The first of these is the age structure of living populations, and the second occurs as a result of attritional death from many causes. When modeled in idealized form, they represent the parent (living-structure) pattern and the first most common derivative death pattern (U-shaped) respectively. One or both will condition all subsequent selection biases of predators and other bone-collecting agents.

Living-Structure and U-Shaped Age Structure Models

Age structure models used in archaeology are based on the frequencies of individuals in a series of consecutive age cohorts of approximately equal duration. In fact, two alternative measures of age composition are used in demographic studies of modern living populations: age frequency-counts and mortality rates (a probabilistic expression of survivorship). Archaeological studies are largely confined to the former because we cannot know the absolute live composition of ancient prey populations. We can only approximate what they were like by extrapolating from the "average condition" of modern unmanaged populations, and by working from an abstract or empirically-defined range of variation for each class of mortality pattern (e.g., Cribb 1987; Lyman 1987b; Stiner 1990b).

When the frequencies of individuals in successive age cohorts are plotted, the *instantaneous* structure of a living population displays a half-pyramidal or staircase form (Fig. 1.1a). Some variation should be expected in the "slope" of the half-pyramid, depending on whether the living population is in a stable, growing or declining state (e.g., Caughley

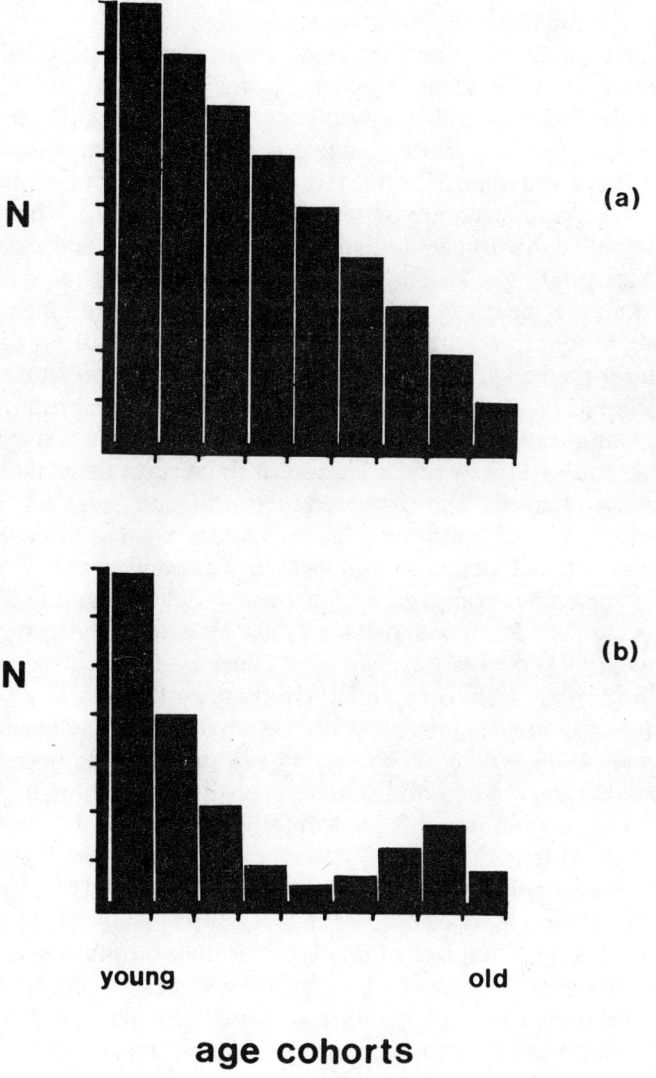

Figure 1.1. Two basic age structure models exemplified using nine age cohorts: (1) idealized structure of a stable living mammalian population; (2) U-shaped mortality model.

1977:121; Sattenspiel and Harpending 1983). The slope varies within predictable limits, however, and this variation can be incorporated with idealized static models for analytical purposes.

If an entire population (or a random sample thereof) is killed off by a sudden event of disastrous proportions, such as fire, volcanic eruption, or floods, the death assemblage will have the same age structure as a living population. This is because all age groups are represented in proportion to their natural abundance, hence the term "living-structure" mortality pattern. The living-structure pattern is sometimes called "nonselective" or "catastrophic" by archaeologists (e.g., Klein 1982a) and paleontologists (Voorhies 1969).

Attritional death in mammals results in a second kind of mortality profile, here termed the U-shaped pattern (Fig. 1.1b). It differs significantly from the age structure of the parent living population, because the probability of dying is significantly lower during the prime years of life (e.g., Caughley 1977:99; Lyman 1987b:126); as a consequence, old animals and especially juveniles appear to be superabundant in attritional death assemblages. The U-shaped mortality pattern reflects age-specific vulnerability to any number of death factors, such as disease, accidents, malnutrition, and many (but not all) types of predators. It is important to realize that, where one agent of attritional death is rare or absent, other factors simply take over and can produce essentially the same result (for a documentary survey of this point, see Stiner 1990b).

The living-structure and U-shaped models were developed by population biologists interested in modern demographic phenomena (e.g., Caughley 1966, 1977). The models have subsequently been borrowed in archaeological studies of ungulate exploitation by humans (e.g., Klein 1978, 1982a, 1982b) and for paleontological studies of how fossil faunas were formed (e.g., Voorhies 1969). In the process, the models have also received new names: "catastrophic" in the place of living-structure, and "attritional" in the place of U-shaped. Unfortunately, these newer labels implicitly suggest causes of death rather than simply describing the age structure patterns themselves. For these reasons, the purely descriptive nomenclature used in population biology is preferred in this book, although corresponding archaeologists' terms are also provided as appropriate.

Some actualistic studies, such as Haynes' (e.g., 1987, 1988) work on proboscideans, indicate that the basic models may not apply equally well to all mammals. The general principles of recruitment and death are the same, but unusual life history and developmental characteristics of certain species introduce some potentially confusing variation in mortality patterns. Like humans, elephants and whales (Savelle and McCartney, this volume) grow relatively slowly, and individuals reach maturity at roughly one third of their maximum potential lifetime.[2] Prolonged development and high levels of parental investment relative to the overall life course bias mortality patterns in unique ways. With the probable exception of cetaceans, the basic age structure models apply for all of the mammals discussed in this volume.

Derivative Patterns

So far, I have described age structures of living and death populations independently of what predators do. From the perspective of modeling human predation, mortality patterns in prey death assemblages represent the *final expressions* of what live populations have to offer, and how humans responded to resources whose existence in no way depended on their own. Predators extract some portion of a living population according to the general search and procurement strategies employed and, in the process, unwittingly create mortality patterns wherever they collect the bones of their prey. However, the relationships between predators and food species are not always directly or simply inferable from the mortality patterns in archaeofaunas.

Prey animals may group themselves in a habitat according to sex and age classes, creating resource "patches" that are biased relative to the prey population as a whole. The structures of death assemblages also depend on whether the predator takes live or dead prey within a given period. Nearly all predators will scavenge (including modern humans, e.g., Gusinde 1961; Bunn *et al.* 1988; O'Connell *et al.* 1988b) and, in these circumstances, they are drawing on sources created by independent agents of death (e.g., other predators or nonviolent causes). The problem of learning how mortality patterns are derived in archaeological cases is further compounded by that fact that those typically created by predators can mimic death patterns occurring in other contexts (Stiner 1990b).

These facts about mortality pose some difficult interpretive challenges to archaeologists interested in human predatory behavior. Coping with these difficulties requires several analytical steps, the first of which is to establish the identity of the bone collectors. This may not be so difficult in Holocene situations, but it can be a serious problem in Upper Pleistocene and earlier archaeological records. Because most mortality patterns are not exclusively diagnostic of cause, taphonomic data rather than age structure are needed to attribute a faunal accumulation to a specific bone collector and to identify preservation biases that might affect the apparent age structure of species therein.

The next analytical step requires a knowledge of the broad array of causes that could potentially produce the observed mortality patterns. Here, one might be dealing with questions of whether prey were scavenged or hunted, or just trying to tease-out more detailed information on hunting tactics. Succeeding presentations will show that some mortality patterns are more diagnostic of procurement strategies than others and that independent contextual data greatly enhance the informative potential of this second stage of analysis.

SCALES OF OBSERVATION

As a group, the papers in this volume provide a clear illustration of how observing single phenomenon — prey mortality — at differing scales can reveal qualitatively distinct kinds of information about human life in the past. The temporal and spatial scales chosen for analysis therefore determine what one can learn, and these scales can and should be adjusted in response to the precise nature of the problem being addressed.

At the finest level, mortality data inform about seasons of resource use with respect to specific places on a landscape. This scale of observation is especially promising for investigating variation within particular subsistence systems and, ultimately, how the internal structure of one culture might have differed from that of another. As Todd's paper illustrates, examining death patterns at the level of seasons and expanding the geographical scale considered also makes it possible to address questions about human territory use. In contrast, mortality profiles representing multiple procurement events over many years are more informative about the *predominant* food search and procurement strategies of human groups. The averaging effect of this intermediate scale has its own distinct analytical value, as it summarizes longer-term interactions between human predators and prey populations. At the grossest level of observation, simultaneous consideration of prey death assemblages widely scattered in time or space provides a measure of human niche, by overriding transitory growth or decline phases of populations and the adjustments that humans normally make to them. When compared to data for nonhuman predators, this scale can be used to monitor the degree and direction of change in prehistoric human foraging adaptations through time.

NUMBER OF AGE COHORTS AND MORTALITY PROFILE CONTOURS

Researchers can scale the age cohort axis in mortality profiles in different ways. For the standard profile representing all age groups in a population, the x-axis typically begins at birth and ends around the maximum potential lifetime (MPL). We all have a pretty good idea when the lifetime begins, but this is no guarantee that we know when it should end (Caughley 1977). In fact, MPL is a rather tenuous parameter by population biologists' standards, yet it is an absolute necessity for constructing and calibrating age structure profiles for death assemblages. The problem is that longevity in a given species can vary extensively between populations as the result of environmental factors, genetics and so on. Record lifetimes for one or a few individuals are an especially poor source for this parameter because they usually involve extraordinary individuals, often living in a protected captive situation (e.g., zoos), and in no way typify maximum longevity in free-ranging populations of the same species. Archaeologists have attempted to cope with the ambiguities inherent in the MPL parameter by using sources for wild populations only and/or by excluding suspiciously long-lived individuals from the control set (e.g., Gifford-Gonzalez, this volume; see also Klein *et al.*'s "ecological longevity" parameter, 1981).

Analysts usually subdivide the age axis into cohorts of roughly equal duration. This way, the data are easily collapsed into grosser age units so that the results of diverse studies can be compared. There is much freedom in choosing how many cohorts to work with; it depends on what you want to know, as well as the sample sizes available for study. However, altering the x-axis scale can have consequences for the overall shape the mortality profile, and, for this reason, subjective comparisons of shape (contour) are not very rigorous analytical procedures. In fact, playing with the x-axis by changing the number of cohorts is not unlike playing with an old rubber band: assuming that you do not break it, the visible texture changes from smooth to bumpy or cracked according to how much the rubber is stretched. More stretching will eventually reveal empty cohorts and, in many cases, an entirely new contour. If prey were born and killed in seasonally restricted periods, waves of similar periodicity may emerge as the age axis is stretched; otherwise the profile may just seem irregular. The nomenclature for mortality profiles discussed above generally applies to systems using somewhere between seven and 15 cohorts, for which the overall contours remain pretty much the same for each type (see Lyman 1987b for a related discussion on sample size).

Still other sources of variation in mortality profile contours are introduced by the aging methods themselves, accidentally via classifications that assume linear rather than curvilinear tooth wear progressions based on crown height (e.g., Gifford-Gonzalez, this volume) or deliberately by dividing MPL into age stages of unequal duration (e.g., Blumenschine, Stiner, this volume). In the latter situation, the lifetime is divided according to life history characteristics that affect the food value of prey and opportunities for procurement by hunters or scavengers.

A WORD ON REFERENCE POPULATIONS

We like to think that biologists who sample wild populations know what they are doing and, indeed, they usually do. However, as archaeologists are aware, so-called "random" samples may be nothing of the kind, and the general problems that archaeologists face in collecting cultural materials are equally critical in demographic studies of animal populations. Because animals move around and may assemble themselves into sex- or age-biased groups, censusing methods must be adjusted to their habits. This requires some knowledge about the behavior of the targeted species to start with and that the census-takers have the logistic capability to carry out the most appropriate procedure. Unfortunately, such luxuries are not always enjoyed by field biologists. Some sampling methods for living populations effectively deal with age- and sex-related differences in threat avoidance, spatial distributions and vegetation density, while others do not. Moreover, censuses may be conducted at vastly different scales than those characterizing human exploitation of prey. Archaeologists want data on prey populations that are independent of human effects, but, in selecting sources of these data, also need to know the temporal and spatial scales of the control sample. We may not be able to avoid inequities between control populations and human-related cases being investigated, but we must be prepared to account for their possible biasing effects.

Finally, there are many research situations in which reliable data on reference populations simply do not exist; a species may be extinct or, if still around, poorly known for a wide variety of reasons. Marine mammals, for example, are difficult to census, and often their existence is so severely threatened that is it is not clear if the present population size and structure is representative or viable for the long term. Species living in tropical forest settings, the last frontiers of relatively unmolested ecosystems, may never be properly censused because they are disap-

Introduction

pearing so quickly. In these circumstances, indirect kinds of control information on parent populations are the only solution. The development of alternative controls calls for considerable ingenuity on the part of analysts (see, for example, Lyman, and Savelle and McCartney, this volume).

NOTES

1. Body size represents an important alternative for aging very large mammals, particularly cetaceans possessing baleen structures in the place of teeth (e.g., Savelle and McCartney, this volume).

2. Here maturity refers strictly to the level of physiological development; that is, when an animal is biologically capable of reproducing. This developmental threshold may or may not correspond to the age at which most individuals actually begin reproducing.

2

Hunting Strategies, Prey Behavior and Mortality Data

George C. Frison

INTRODUCTION

Hunting became a major means of subsistence since the time humans acquired the technology to manufacture weaponry with which they could predictably kill animals once they were outmaneuvered. In the process, hunting became a high prestige activity confined primarily to males, and weaponry became a focus of attention beyond its functional aspects. Hunting is hard work, as are both the efforts that must go into the manufacture and maintenance of weaponry and processing the products of the hunt. The latter apparently became an important female activity, if our archaeological and ethnological data are correct. In critical situations such as mass animal kills, however, the need to prevent loss of meat may have temporarily overridden a strict sexual division of labor.

Much of what has been said about prehistoric hunting lacks credibility because the observers were untrained with regard to hunting and the behavior of prey. In fact, few practicing anthropologists are sufficiently versed in either area of study to properly analyze what actually goes on in a hunting situation. Some anthropologists also tend to depreciate experimental hunting with aboriginal weaponry because the investigator cannot place himself (or herself) in the footwear of the prehistoric hunter. These criticisms are valid to some extent, but, in refusing to seriously consider information that can be obtained from the actual experience of hunting and from people who are intimately familiar with the behavior of wild species, some potentially useful applications for archaeological and ethnological studies are lost.

In North America, there are few reliable pictorial records of prehistoric hunting with the possible exception of some painted and pecked pictures on stone, and, more rarely, on perishable media such as wood or hide. Pictorial records can be informative as to the kind of weaponry used and the animals that were taken, but I know of no instance where they accurately portray details of a procurement method or the sequence of events. There are a few descriptions of native American hunting by early European explorers, fur trappers, and others, but most of these also lack the detail needed to understand the procurement tactics used, probably because not all of them were experienced hunters. Consequently, textbooks, museum exhibits, and other interpretive displays rely heavily upon the depictions of individuals who lacked familiarity with the real contingencies of animal procurement, such as the limitations imposed by the behavior of the species involved, the size and composition of the hunting party, the time of year, weather, vegetative cover and topography, and numerous other factors.

LASTING IMPRESSIONS

Gladwin (1947:97) portrayed a bison "jumping" event in which three hunters, one waving a blanket, one waving his arms and shouting, and another carrying a spear, are chasing eight bison (Fig. 2.1). Four bison have leaped, without hesitation, over a precipice and the others blindly follow. To those unfamiliar with bison, this scene may evoke no further comment, but to anyone who has spent time around this species, it is ridiculous and contrary to normal bison behavior. Unfortunately, these kinds of spectacular pictorial interpretations are too often regarded as historical fact and, in turn, have resulted in erroneous interpretations about human subsistence strategies in historic and prehistoric situations.

In the early 1960's, a documentary film was released entitled *The Hunters*. The film recorded a hunting trip in which a group of Bushmen wounded and subsequently pursued a giraffe. Thus well-known film has become a standard part of introductory anthropology courses, although, in reality, it is a portrayal of animal procurement in which the hunters *mismanaged* their initial contact with the animal and inflicted an apparent flesh or stomach wound. Instead of falling back and allowing the animal to lie down, the hunters pursued it too closely, resulting in several days of long distance trailing before the animal was finally cornered. At this

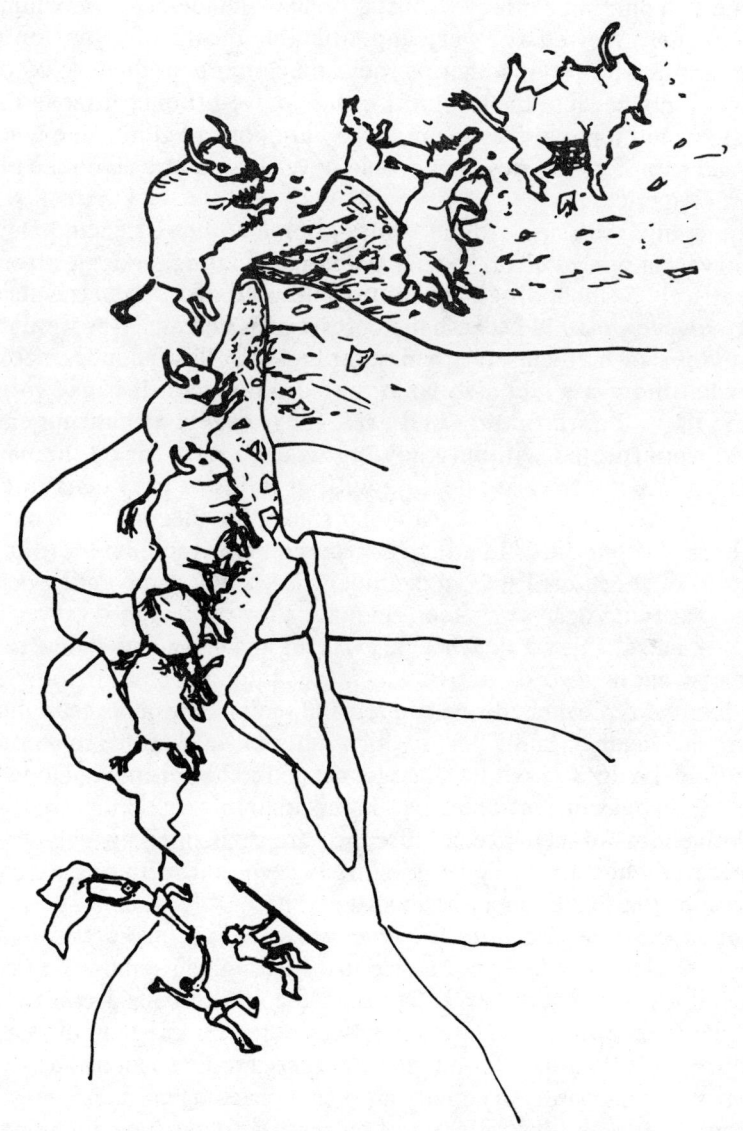

Figure 2.1. The way not to jump bison (drawing from Gladwin 1947).

stage, someone took pity on the giraffe and downed it with a rifle so as not to prolong its agony unnecessarily. The events in this epic film are imprinted on the minds of too many archaeologists and ethnologists as the epitome of a hunting strategy. Thus, it is easy to understand why hunting is erroneously viewed as a very unpredictable means of provisioning a human group with food; gathering looks much more predictable by comparison if only because the resources cannot move-off once discovered. If prehistoric hunters were as inept as they are portrayed in some popular films and explorers' diaries, they would never have survived on the plains of North America.

The companion book to the film, *Man the Hunter* (Lee and Devore 1968), was an honest effort to analyze human hunting strategies from an more strictly anthropological point of view. It also has some serious flaws, however, mainly because the observations and analyses were made by individuals who themselves had never hunted. This fact influenced not only what they saw, but also what they did not see. It is not valid to assume that all anthropologically relevant aspects of hunting can be properly interpreted without knowing what to look for on the part of humans and what they are dealing with in terms of prey behavior and hunting conditions. From this particular study, the student might conclude that there was *one ideal* hunting strategy and that each hunter followed some sort of check-off list from the time when he left camp until when the animal was lying dead at his feet. In reality, the experienced hunter need only look outside his abode when he awakens and consider what he sees in terms of what he already knows about the animal species he intends to hunt; i.e., the typical condition of prey and their movements according to season, the weather, and a host of other determinants. If the hunter takes to the field, he does so with a strategy mapped-out in his mind, along with some alternatives in case one fails. In this kind of situation, prey animals stand much less of a chance because they are challenged by a hunter who can predict what they will do to a large extent and who possesses and knows how to use the proper weaponry.

On the positive side, *Man The Hunter* inspired archaeologists to take a serious look at their data and realize that hunting was a more significant and complex segment of human behavior than it had been accorded up to that time. It is clear that hunting practices deserve more than the superficial coverage previously found in most ethnographic accounts, as well as in interpretations based on activity areas in archaeological sites.

Some textbook illustrations and museum displays have disseminated even more misinformation on *prehistoric* hunting. Mammoth procurement in particular has captured the popular imagination and, in most cases, portrayals violate all rules of intelligent, successful hunting. Too many investigators continue to believe that prehistoric hunters usually forced large animals into swamps, over cliffs or toward any large obstacle in order to impede their progress and allow shouting hunters (accompanied by barking dogs) to spear them. There is usually a tragic element too: a member of the hunting group is killed or maimed in the process.

Trapping large prey in bogs is a common scenario that fails to recognize the weight of the target species. A dead mammoth weighing anywhere from 7,000 pounds to twice that, or a large bison weighing a ton or more, would be next to impossible to extract from a bog. There is the additional problem of butchering a large enmired carcass, not to mention protecting the edibility of the meat products from the encroaching mud. A more prudent strategy would have been to first drive the animal out of the bog and kill it on solid ground, a procedure that many cultures possessed the knowledge, manpower and technology to perform.

On the other hand, most large land mammals, including elephants and bison, are greatly attracted to swampy areas and have no trouble moving in and out of them (Fig. 2.2). Animals that become trapped in bogs are usually sick, crippled or old and are among the least desirable as food. During a decade of intense study of elephants in Wankie National Park in Zimbabwe, for example, Gary Haynes (personal communication) claims never to have encountered an individual caught in a bog unless it was starved or crippled. Conversely, animals that regularly frequent bogs can be difficult or impossible to drive or otherwise entice into a bog if they decide not to go.

Depictions of "jumping" large prey over cliffs are every bit as erroneous. Scenes of horses or bison running from persons shouting and waving objects and, of their own free will, leaping over precipices to their death can be found in literature on hunting in the Old World and the New World alike (e.g., Fig. 2.1). We now know that jumping bison requires very specific conditions in order to succeed: the herd must be closely-packed, of sufficient size, and moving fast enough so that the animals in the rear will be unaware of the sudden drop in the terrain ahead and force those in the lead over the edge. A small group (such as 6-10 animals) is almost impossible to drive over a bluff because the animals have time to perceive the danger, and they immediately react by swerving or reversing direction to avoid it. While bison may appear ungainly in their movements, they can change their direction as quickly or more quickly than any animal of equivalent size (see Fig. 2.3).

These imaginative depictions also show a total disregard for human life: mammoth kills are usually pictured with a dead and/or crippled

Figure 2.2. Domestic cattle in bogs. The cow in the center is up to her belly in mud and grazing contentedly. Her calf is with her (a), and later jumped out of the bog (b).

Figure 2.3. Decision-making time while handling an unhappy bison cow: a split-second later she jumped over the author and splintered a section of the corral fence.

hunter being dragged from the scene. In reality, band-level societies could ill afford the loss of hunters through careless behavior. Hunting large mammals is an occupation during which there is a higher than normal potential for bodily injury. Any injury that affects sight, hearing and reaction time, can negatively affect the probability of success. Frost bite, for example, can cause serious lifetime impairment. Even a minor disability from injury or exposure would impair the ability of the hunter and jeopardize the future of himself and his relations. Hunting requires that all human facilities be acute, and decisions are often conservative to avoid losing more than can be gained.

As killing techniques go, bear hunting has been subjected to the most ridiculous representations. Not least of these is the familiar teddy bear cowering in a corner as human hunters strike with clubs, rocks and spears. Anyone who has experienced the aggressiveness of a bear, particularly when cornered and/or wounded, or observed a grizzly snap an animal's neck, or witnessed the speed with which they cover distances will corroborate the totally ridiculous nature of this kind of situation. Renditions in which prehistoric hunters hold large stones in both hands above their heads, ready to fling them from a distance at large animals such as mammoths, bear and bison, are nearly universal. In reality, this would be a totally ineffective way of killing or wounding the animal unless it were intended as the final *coup de grace* at close range and the animal was already down and mortally wounded.

Much ado has also been made of the use of fire for driving animals in a particular direction. Driving animals, wild species in particular, requires split-second decisions and reactions in direct response to the movements of the animals. Any time a fire is lit, whether in grass, brush or forest, the vital potential for instant reaction and control is lost. Burning vegetation will attract grazing animals to the new growth once it appears, but attempting to predictably drive animals with fire into a corral, natural trap or over a precipice will not succeed except in extremely rare cases where the wind conditions, herd location and landforms happen to be just right.

PROCUREMENT OF WILD NORTH AMERICAN MAMMALS

Bison, pronghorn antelope, mountain sheep, and probably Pleistocene mammoth, were amenable to communal procurement methods. They are also the most common species found in High Plains archaeological sites. Each modern species displays a distinct set of behavioral characteristics, and hunters had to know how to exploit their behavior for procurement events to be successful. Numerous archaeological sites on the Plains attest to their past successes in these endeavors.

Prey behavior was a strong determinant in any hunting episode. Bison, for example, were successfully maneuvered into a variety of natural and artificial features, including arroyo head-cuts, parabolic sand dunes, corrals at the base of perpendicular drops, steep talus slopes or artificial ramps, and drive lanes formed by fences terminating at corrals. Although appropriate for bison, the same features would not work for trapping pronghorn, deer or mountain sheep. Mammoths may have been taken in natural features such as arroyo traps, although the North American mammoth kill sites presently known usually do not preserve this kind of information. One exception is the Colby site (Frison and Todd 1986), where the landforms present at the time of the kills can be reconstructed to some extent. In this case, a steep-walled arroyo may have been used to trap animals, although it cannot be proved.

Pronghorn were regularly taken in traps, but in types that would be useless for deer, elk or bison. One kind of pronghorn trap consisted of a drive line leading into a circular corral. The entire arrangement is made of brush (see Frison 1978b:254-255), which pronghorn will refuse to jump over due to a peculiar facet of their behavior. In contrast, deer and mountain sheep would not hesitate to jump the same fence.

Mountain sheep display a different set of behavioral characteristics, and it is relatively easy to maneuver them into artificial traps. Knowing this has allowed investigators to identify a large number of wooden structures in the mountains of northwest Wyoming and adjacent areas of Idaho and Montana as mountain sheep traps. Like their domestic relatives, mountain sheep can be driven between converging fences, up ramps and into catch pens. Some special conditions of fence construction are required, however, to compensate for sheep's exceptional ability to climb and jump. Except for the rutting period in late fall, rams tend to congregate in groups away from the nursery herds and display very different behavior. Consequently, the nursery herds require different procurement strategies than ram herds, and both types of traps have been recorded archaeologically (see Frison 1987).

There is evidence that Mountain sheep were taken in nets as early as 9,000 years ago in the Absaroka mountains in northwest Wyoming (Frison et al. 1986). The natural reactions of mountain sheep when netted must be understood in order to properly interpret the use of nets in prehistoric sheep procurement. Unlike deer and pronghorn, which go totally berserk under nets and rip them to shreds, mountain sheep will struggle for a short time and then become docile (Fig. 2.4). It is a relatively simple matter to

Figure 2.4. Mountain sheep under a net. Note the ewes standing relatively calmly before the next part of the operation.

drive sheep into a net and kill them with clubs. The same strategy would be difficult if not impossible for trapping deer or pronghorn. I do not mean to imply here that deer were never taken in nets; rather, a different kind of netting strategy is needed for deer.

The mammoth no longer exists, and its behavioral characters cannot be directly observed. However, if mammoth social habits and herd structure were organized in the same way as those of modern African elephants (Fig. 2.5), it would have been difficult for Clovis hunters to face a matriarch, much less dispatch her and her entire family, given the limitations of their weaponry. Taphonomic and archaeozoologic data on North American cultural mammoth kills are not yet sufficient to determine season of death, herd structure, and other conditions relevant to mammoth procurement. However, observations on African elephants suggest that taking individual animals that stray away from the immediate protection of the matriarch is quite feasible (Frison 1989). One or more knowledgeable hunters could carefully follow a herd of animals and locate one that has strayed far enough away from the others. By allowing for the animal's relatively poor eyesight but highly-developed sense of smell, the hunters could get close enough to inflict a wound before the matriarch was able to focus on the source of the danger. The wounded animal would gradually drift away from the herd to a place where it could be killed. This kind of mammoth hunting is the antithesis to the dramas portrayed in popular texts and museum displays; there are no bogged animals, barking dogs, shouting hunters or maimed companions.

TAPHONOMIC AND AGE STRUCTURE STUDIES

A large number of archaeological bonebeds representing large-scale, communal kills have been identified and investigated in recent years. Because mass deaths represent short-term events, seasonality is very pertinent to understanding the procurement context. In many cases, seasonality determinations are possible in mass death sites because many individual prey animals are represented. Determining the correct time of year of animal procurement operations is vital to an accurate understanding of prehistoric human hunting methods. Animal behavior changes rapidly, not only with the seasons, but also with the time of day, weather, feeding conditions, animal nutrition, age and sex.

Three decades ago, recovery and interpretive methods for dealing with bonebeds had not been developed and, consequently, much of the data base was improperly curated, misidentified and/or discarded, as in the case

26

Figure 2.5. African elephant family herd at a watering hole with the matriarch and sub-matriarch standing guard.

of the first excavations at the Agate Basin site (Roberts 1943). Perhaps the earliest efforts alerting archaeologists to the scientific potential of culturally-associated bonebeds are to be found in Kehoe's (1967) work at the Boarding School Bison Kill, Wheat's (1967) analysis of the Olsen-Chubbuck site, and Lorrain's (1968) analysis of Bonfire Shelter. Lorrain was able to construct general mortality patterns for the bison remains at Bonfire Shelter, but not the seasons of death. Wheat postulated an early spring kill, which later proved to be incorrect; subsequent studies using new, more accurate aging techniques for bison calves, showed that the animals were killed in late summer (see Frison 1978a:46). The erroneous seasonality determinations led to still other errors in interpretation concerning the procurement situation involved, the mode of operation and of the butchering, processing and utilization of the meat products.

Credit for the earliest applications of mortality studies to archaeological assemblages must be accorded to the late Dr. Paul McGrew, then the paleontologist at the University of Wyoming, who encouraged and instructed numerous researchers in their work on mass death assemblages. The first successful applications of animal population data to archaeological interpretation in North American sites were by Reher on a series of bison kills (1970, 1973, 1974) and by Nimmo (1971) on a pronghorn assemblage. These studies followed the methodology developed by Voorhies (1969), who formerly applied them to a paleontological assemblage of Pliocene age.

The method for establishing the age structures and seasonality of animal kills in these cases was based on tooth eruption and wear, and relied on adequate samples from catastrophic kills of bison and pronghorn. The precision of the method was demonstrated later on a mule deer sample (Simpson 1984). This method uses only young animals (preferably newborns and individuals up to two years of age), for which tooth eruption can be measured accurately enough to assign age in months. In the case of bison, it soon became apparent that domestic cattle matured on a different schedule than bison (see Frison and Reher 1970), and adjustments have since been made to compensate for the differences in development rates.

GEOARCHAEOLOGICAL EVIDENCE AND
ANIMAL BEHAVIOR

In many cases, geoarchaeological data can establish the landforms used in procurement operations even if they are no longer visible. Land form types are a critical consideration when maneuvering animals, since prey may be more willing to move into one type of feature than another. There are certain features into which animals will simply refuse to go. Mountain sheep, for example, prefer to move up steep slopes when disturbed, whereas bison and domesticated cattle usually move better in drive lanes that progress slightly upslope rather than downslope. They will also move better along a curved drive lane than one that is straight, as is clear at the Ruby bison corral site (Frison 1971). Prehistoric bison hunters on the northern High Plains undoubtedly were aware of this facet of bison behavior as well. Through a comprehensive understanding of both landforms and animal behavior, better interpretations of prehistoric procurement situations may be realized.

CONCLUSION

Mortality data obtained from the study of animal kill sites on the northwestern High Plains and in the central Rocky Mountains have been used to interpret hunting strategies for large animals. Most of the communal kills display catastrophic age structures, but this is not true of all assemblages, and bonebeds do not always represent the actual place of the kill. Mortality data are amenable to quantitative analyses and can reveal much about prey populations and seasons of procurement by humans, but their utility as indicators of the actual methods used to maneuver large animals into natural or artificial traps is more limited, particularly if considered in isolation.

Archaeologists too often are reluctant to utilize animal behavior studies or the direct experiences of hunting for analyzing past hunting methods and the limitations imposed by prey species. Instead, many archaeologists continue to rely upon the observations of non-hunters and on artists' misconceptions of animal behavior and hunting methods. As a result, many analyses of prehistoric subsistence strategies involving large mammals are incorrect. The ways that archaeologists most often conceive of the hunting process and the strategies used require revision if archaeologists are to accurately reconstruct past cultural systems of hunting peoples.

Animal behavior is a diverse and complex topic. Procurement of one species as opposed to another, or particular individuals within a species according to sex or age, may affect hunters' procurement and processing decisions in important ways. The idea that the same hunting strategy can be applied to all prey species is erroneous and results in faulty interpretation. Human hunting behavior can be better understood through more

thorough use of the many different sources already at hand. These include considerations of the prey species involved, hunting group size, quantity of meat products and their condition, and butchering and processing. When data from alternative sources fail to fit into the overall picture of a procurement event, this may be an important signal to the investigator that careful reevaluation of the behavioral criteria for interpretation is needed.

3

Taphonomy and Early Hominid Behavior: Problems in Distinguishing Cultural and Non-Cultural Agents

John D. Speth

INTRODUCTION

Three decades ago, taphonomy was practically unknown to archaeologists. Today, taphonomic studies have become absolutely central to our understanding of hominid origins and the role of hunting and scavenging in early hominid subsistence strategies, and the importance of taphonomy in studies of later human evolution is becoming increasingly evident (e.g., Behrensmeyer 1975, 1987; Behrensmeyer and Hill 1980; Binford 1981, 1984; Blumenschine 1987; Brain 1981; Bunn 1986; Bunn and Kroll 1986; Fisher 1987; Haynes 1988b; Klein 1987, 1989; Lyman 1985; Marshall 1986; O'Connell *et al.* 1988a, 1988b; Potts 1983, 1988; Shipman 1981, 1986b; Todd 1987). Taphonomy has rapidly emerged from the level of interesting cautionary tales to a full-fledged discipline in its own right. Thanks largely to these investigations, we have come to realize that patterning in the Plio/Pleistocene archaeological record, which until recently we have simply assumed to be a direct reflection of hominid behavior, may often be largely or entirely the product of nonhuman agents (Behrensmeyer 1987; Binford 1981; Brain 1981). This realization has forced us to reevaluate many of our traditional interpretations of pre-modern human behavior and, in a growing number of cases, has led us to strikingly different conclusions.

It is not my purpose here to review the many important contributions that taphonomy has made to early hominid archaeology; these are ably summarized in a number of recent reviews (e.g., Behrensmeyer 1987; Binford 1981, 1985; Bunn and Kroll 1986; Potts 1988). Instead, I briefly consider a few broader issues of a taphonomic nature that I think have not received as much attention as they deserve. What concerns me particularly is the fact that, while we have made tremendous strides toward recognizing the importance and complexity of taphonomic agents in Plio/Pleistocene (and later) site formation (e.g., fluvial transport and sorting, carnivore transport and attrition, decay, trampling), we have not made as much progress in recognizing the high degree of behavioral variability that is likely to characterize both early hominid and later human use of animal resources. Thus, while our rapidly increasing sophistication in taphonomic modeling provides an essential and long overdue corrective for the often naive reconstructions that we made only a few decades ago, we now run a new and potentially equally serious risk of falsely attributing to natural causes important aspects of the patterning in the archaeological record that in fact may be the product of the very human behavior we seek to identify and explain. The use of "multiple working hypotheses," a strategy often advocated by scholars working with the early hominid archaeological record (e.g., Isaac 1983) does not necessarily alleviate this problem, simply because so many of the competing hypotheses that are presently at hand relate to anticipated taphonomic rather than behavioral variability.

Fortunately, recent and ongoing studies by archaeologists, ethnologists, and behavioral ecologists among living foragers (e.g., Binford 1978; Bunn et al. 1988; Hill et al. 1987; O'Connell et al. 1988a, 1988b; O'Connell and Marshall 1989; Yellen 1977, see also Alvard and Kaplan, Hudson, this volume) and among small-scale subsistence horticulturalists (e.g., Beckerman and Sussenbach 1983; Dwyer 1974; Hames and Vickers 1982; Marks 1976; Sponsel 1981) are beginning to redress the present imbalance between our models and reality, both by documenting the tremendous behavioral variability that characterizes contemporary human use of animal resources and by systematically exploring the nutritional, ecological, and sociocultural factors that give rise to this variability.

Here, I explore some specific aspects of this behavioral variability, highlighting a few specific areas where I believe we are in particular danger of falsely ascribing patterning in the early hominid (and later) archaeological record to natural (i.e., nonhuman) rather than human agencies. I should hasten to point out that in the following examples I do not mean to imply that current taphonomic explanations are necessarily wrong, only that we must exercise as much caution in ascribing a pattern to taphonomic causes as we do in attributing it to human causes. Given the overall focus of this volume on mammalian mortality patterns, I will include examples that specifically involve the age and sex composition of prehistoric faunal assemblages.

BONE FRAGMENTATION

We often assume that humanly-produced Plio/Pleistocene (and later) bone assemblages will be highly fragmented, presumably because contemporary human foragers, who often provide our models of the past, normally utilize all edible portions of a carcass intensively, including the lipids in the marrow and cancellous tissue of bones (e.g., Brain 1981:53-54): Bunn et al. (1988:443) note "the routine Hadza practice of chopping and boiling these parts [axial carcass units] for grease"; Yellen (1977: 291) emphasizes the fact that the !Kung San almost always smash up bones so that they can be placed in cooking pots and boiled to extract the grease. By contrast, Plio/Pleistocene and later Pleistocene faunal assemblages that are not highly fragmented are often suspected of being the remains of natural deaths or carnivore scavenging activities (e.g., Brain 1981). The underlying assumption, of course, is that, if hominids had been involved in processing the carcass, they would have broken the bones apart to extract the lipids like modern foragers.

There are two potential problems with this assumption (see also Potts 1988:116-118). First, as far as we know, hominids as recently as the Late Pleistocene lacked the technology (i.e., highly controlled use of fire, water-proof containers, stone-boiling) needed to effectively boil grease out of the cancellous tissues of limb epiphyses and vertebrae. This eliminates one of the major reasons that Holocene and contemporary foragers routinely smash up bones (Binford 1978).

Second, the deposits of fat in the marrow provide a critical energy reserve for the living animal, which may be mobilized under conditions of nutritional stress and become partially or largely depleted (Speth 1983). Most studies of early hominid subsistence are set in African environments. African ungulates are subject to nutritional stress each dry season, so much so, in fact, that many subsist on submaintenance energy budgets for up to three or four months every year (Sinclair 1975). Since most African ungulates have very little body fat to start with, by the end of the dry season many of them have used-up their fat reserves, including the lipids in the marrow (e.g., Ledger 1968; Sinclair and Duncan 1972; Brooks et al.

1977; Brooks 1978). Smashing bones to extract fat from the marrow at this time of year, therefore, may not always have been a rewarding activity for early hominids. This would be especially true in situations where they were scavenging carcasses that had died of natural causes, since many of these deaths would have been linked directly or indirectly to nutritional stress. Thus, the degree of bone fragmentation may at times reflect seasonal differences in hominid use of animal resources, rather than the extent to which the faunal remains were the products of human as opposed to nonhuman agents.

Factors other than seasonal resource stress may also affect the physiological condition of an animal and hence the intensity with which its bones are likely to be processed for marrow fat by human hunters or scavengers (Speth 1983, 1987, 1989). Calves, for example, typically have very limited body-fat reserves throughout the year and may often be ignored by human hunters or scavengers (see Speth 1983:95, Grønnow 1987:146, and Stiner, this volume, for archaeological examples). Similarly, female animals carrying a full-term fetus, or that have just calved and are nursing, may also be fat-depleted, while adult males and non-reproductive females may be in better shape and hence richer sources of fat. Likewise, very old individuals of both sexes typically have smaller total fat reserves than prime-aged adults and may also be ignored in favor of animals in better condition. Finally, males that actively participate in the rut in a given year often stop eating and may expend much of their fat reserves, whereas younger or older males that do not enter the rut that year are likely to remain in better condition. In sum, animal condition is a complex phenomenon which varies along a number of different dimensions. And since the physiological condition and nutritional needs of the human foragers probably also vary in a complex manner, the extent to which animal bones will be fragmented to retrieve marrow fat or bone grease is unlikely to be uniformly high in all human situations.

Bearing in mind the many sources of variability just discussed, it is quite possible that the degree of trabecular bone fragmentation (e.g., spongy bone of the axial skeleton and limb epiphyses) by humans may have increased rather suddenly and dramatically toward the end of the last glaciation, when the technology for bone boiling presumably first became available (Speth 1989). Thus, the degree of bone fragmentation may be more reliable as an indicator of human activity in the Late Pleistocene and Holocene than during earlier phases of human evolution. This observation raises an interesting issue concerning early hominid transport of vertebrae from kills or scavenged natural death sites to central places or "basecamps." The conventional view is that axial elements, particularly from larger mammals, will normally be stripped of edible tissue and left behind at the death site, while meatier, marrow-rich appendicular elements are more likely to be transported away (see discussion and references in O'Connell et al. 1988a:142ff). Thus, the presence of high ratios of appendicular to axial elements in archaeological contexts such as Olduvai Gorge (Bunn and Kroll 1986; Potts 1988), together with cutmarks or impact fractures, have been central to arguments that early hominids regularly transported meat and marrow bones to central places.

O'Connell et al. (1988a), however, have recently observed that the Hadza, contemporary foragers in Tanzania, frequently transport vertebrae away from kill or natural death sites, even those from animals as large as giraffes. In surveying the ethnographic literature, they note similar transport patterns among other contemporary foragers, including the !Kung San in Botswana (Yellen 1977) and the Nunamiut Eskimo in Alaska (Binford 1978). On the basis of these observations, they challenge the conventional wisdom concerning discard of axial parts at procurement sites and suggest that previous interpretations of the early hominid faunal record, which rely heavily on limb to axial element ratios, are in need of reevaluation.

O'Connell et al.'s (1988a) observations concerning the transport of axial elements are both interesting and provocative. But should we reject the conventional view that axial elements of larger animals are seldom transported in other human/hominid contexts, and replace it with a new, programmatic assumption that such elements are commonly transported? It is worthwhile to first briefly consider why axial elements might be transported in the first place. In contemporary settings, for example among the Nunamiut (Binford 1978), hunters often dry the vertebrae as articulated units rather than spend time removing all of the edible tissue from these highly irregular bones. Another reason for transporting axial elements that may be even more important in the case of large mammalian prey is to extract the lipids from the cancellous tissues (Binford 1978). These lipids may be nutritionally and calorically vital for hunter-gatherers facing recurrent periods of food shortage (Speth 1987, 1989, 1990). But these lipids are difficult and time-consuming to extract and, among contemporary foragers, are generally obtained by smashing and pounding the bones and then boiling them (Yellen 1977:291; Bunn et al. 1988:443). Since the technology to accomplish this may not have become available until the latter portion of the last glaciation, foragers of the Late Pleistocene and Holocene may have been much more inclined, at least seasonally, to transport the vertebrae of large mammals than were their premodern hominid predecessors. Thus, any sort of fixed or normative standard concerning human proclivities to transport axial elements may be misleading.

PROPORTIONAL REPRESENTATION OF SKELETAL ELEMENTS

Evidence of early hominid activity frequently occurs in or adjacent to ancient stream channels, in deltaic deposits, or along lake shores (Butzer 1978:209). Understanding the role of fluvial and lacustrine processes in the formation of these sites therefore is pivotal for drawing reliable behavioral inferences from Middle Pleistocene and earlier archaeological records. Based on flume experiments, Voorhies (1969), and later Behrensmeyer (1975), classified skeletal elements into a series of dispersal groups, based on size, shape, and density (see also Shipman 1981:31-38). Skulls and mandibles belong to the least transportable group, whereas ribs, vertebrae, the sacrum, and the sternum belong to the most transportable group. Most of the limb elements fall within an intermediate transport group. The dispersal potential of specific elements depends to some extent on the size of the animal involved and on the manner and extent to which the bones were broken beforehand. These dispersal groups provide invaluable clues to whether an assemblage represents a transported or lag deposit, as well as the degree to which a fossil bone assemblage has been sorted by moving water.

Unfortunately, interpretations of skeletal element frequencies in Plio/Pleistocene faunal assemblages based on Voorhies' dispersal groups may not be as straightforward as we might wish. Binford (1978:74) developed a series of utility indices (the so-called "Modified General Utility Index" or MGUI) for various anatomical parts of ungulates, based on the amount of meat, marrow, and grease each part contains. Of importance here is the fact that there is a high degree of correspondence between membership in Voorhies' high and low dispersal groups and general food utility: parts with high dispersal potential also tend to be parts with high general utility (e.g., ribs, vertebrae, sternum), or they are small elements likely to be transported by humans as riders attached to high utility parts (e.g., patella, carpals, tarsals). Many of the elements with intermediate dispersal potential (e.g., femur, tibia, humerus) also have moderate to high utility values. Transport of these skeletal elements by prehistoric humans is less predictable overall and probably depended on whether carcass units were muscle-stripped and processed for marrow at the kill and discarded there, or instead were transported elsewhere for further processing.

Contemporary foragers (and perhaps early hominids) commonly intercept and kill animals, or scavenge dead ones, in and adjacent to drainages where forage and water are available. Neither the fact that the element frequencies match those predicted by Voorhies dispersal groups, nor the fact that the bones are found in a fluvial depositional context, in and of themselves demonstrate that taphonomic rather than hominid agencies are responsible for the assemblage.

Element frequencies can also reflect the differential destruction of softer, less dense skeletal elements by attritional processes, including natural decay, gnawing by predators or trampling. The most easily-destroyed bones tend to be those which are the least compact and lowest in bulk density (Grayson 1989; Klein 1989; Lyman 1984, 1985). Under-representation of low-density elements need not be a sign of the operation of nonhuman agents, however. As shown by Lyman (1985; see also Klein 1989:378), there is a weak negative correlation, between Binford's Modified General Utility Index and bulk density, based on measurements of deer bones. In other words, elements that are most susceptible to loss or decay through carnivore attrition or other natural processes also happen to be parts of high utility which are likely to be transported away by hominids. Metcalfe and Jones (1988) have recently proposed a simpler version of Binford's MGUI index, called the Food Utility Index (FUI). The FUI shows a somewhat clearer, though still weak, negative correlation with bulk density. If one focuses only on the grease index, rather than on composite indices of general utility such as the MGUI or FUI, the negative correlation with bone density becomes highly significant (Brink and Dawe 1989:136). In other words, porous, low-density elements that are most susceptible to decay and other attritional processes are also among the richest in grease and therefore the ones most likely to be smashed-up by the site's inhabitants during grease-rendering.

One of the principal reasons that early hominids transported bones to central places or "habitation" sites may have been for their content of marrow fat. Thus, the relationship between bulk density and Binford's (1978:27) marrow utility index is of considerable interest. The higher the bulk density of the bone, the higher its utility as a source of marrow (Pearson's r is positive and significant, $p<0.01$). This means that a strong bias in anatomical representation in favor of bones resistant to attrition or decay may actually reflect hominid selection for bones rich in marrow rather than differential preservation of the densest elements.

Seasonality obviously plays a key role in determining the fat content of ungulate marrow bones (Speth 1983). During periods of nutritional stress, such as the dry season in many tropical and subtropical regions, marrow fat is depleted in an orderly sequence, beginning in the proximal limb elements and progressing distally. In the one African ungulate — the impala (*Aepyceros melampus*) — for which the rate of fat depletion within particular limb elements has been carefully documented, fat values in the proximal limbs (i.e., humerus and femur) were either above 70% or below 50%, with very few intermediate values (Reich 1981). This implies that impala were almost always either in good condition or in poor condition, and that once an animal's condition began to decline, the fat in the marrow of the humerus and femur were depleted rapidly. Thus, high proportions of distal limb elements in a site, commonly taken as a signature of hominid scavenging, may instead reflect the season of procurement rather than the method by which the hominids procured the prey.

Biases toward higher proportions of front versus rear limb elements have often been taken as a sign of scavenging, based on the assumption that primary predators such as lions or hyenas will devour the meatier rear limbs first, and abandon (at least in larger prey) more of the lower-utility front limbs to scavengers (e.g., Blumenschine 1986a:643). However, in the few ungulate species where marrow depletion sequences have been studied in detail (e.g., moose, roe deer, impala), there are differences in the rate at which fat is mobilized in the marrow of the front and rear limbs, particularly in the more distal skeletal elements (Peterson *et al.* 1982; Ratcliffe 1980; Reich 1981). Specifically, fat tends to be depleted sooner in the metatarsals than in the metacarpals. Too few studies have been conducted to be sure whether this pattern is widespread in African or other ungulates or if the difference in depletion rates is large enough to be nutritionally significant (see, for example, Davis *et al.* 1987 for data on a small caribou sample indicating no significant differences in rates of marrow fat depletion between bones of the front and rear limbs). If future research confirms the importance and wide-spread occurrence of this phenomenon, we might expect hominid occupations that occurred during the most stressful portions of the dry season to show some bias toward distal front limb elements, regardless of how the animals were procured. Of course, should the patterns of fat depletion differ among prey species, as is suggested in a brief review by Fuller *et al.* (1986), the ratio of front to rear limbs in archaeological sites should vary accordingly.

PROPORTION OF JUVENILES AND ADULTS

As a final example, let me comment briefly on criteria proposed by Vrba (1976, 1980) to distinguish scavenging from hunting. According to

Vrba, a high percentage of juveniles in a bone assemblage is indicative of hominid hunting, and this should be increasingly true as the body size of prey increases. Scavenging, on the other hand, is thought to be indicated by a low percentage of juveniles, based on the assumption that juveniles killed by carnivores would be largely or entirely consumed before hominids would have had access to their remains. Recent field studies of carcass destruction by hyenas in the Serengeti grasslands of East Africa lend considerable support to these arguments (Blumenschine 1987, also this volume).

While Vrba's specific conclusions concerning the faunal assemblages from South African Plio/Pleistocene cave deposits may be entirely correct, her argument assumes a rather static and normative pattern of human faunal exploitation which may be misleading if applied as a universal to the early hominid (or later) archaeological record. The carcasses of immature animals, especially calves, contain very little body fat (Speth 1983). In the youngest age classes, even the marrow tissue may be nearly devoid of lipids and these deposits are very sensitive to depletion. Young animals therefore provide a relatively poor source of energy for foragers and, during seasons when alternative lipid- or carbohydrate-rich resources are available, their exploitation may be minimized or avoided altogether (see also discussion in Stiner 1990b:316-317). Thus, a low percentage of juveniles in Plio/Pleistocene (or later) assemblages may have little or nothing to do with conditions of bone preservation or whether meat was procured by hunting or scavenging, but instead may be indicative of the season during which foraging activities took place.

CONCLUSIONS

We are making great strides toward recognizing the importance and complexity of nonhuman agents in site formation, but we have not made as much progress in anticipating and modeling the kinds of variability that may have characterized hominid use of animal resources. Thus, while taphonomic studies have provided a valuable corrective for the often naive behavioral scenarios of two or three decades ago, the pendulum may be swinging too far too fast in the opposite direction, thereby increasing the likelihood that we now will falsely attribute to nonhuman causes the very behavioral patterning in the archaeological record that we seek to identify and explain. The recent upsurge of interest in the hunting, butchering and processing strategies of contemporary foragers and small-scale horticulturalists aids in redressing this imbalance. When combined with more detailed considerations of prey physiology and metabolism, these studies are vital to developing appropriate models for the archaeological record. Until we can explain the tremendous behavioral variability we see today, we will be severely handicapped in our attempts to correctly interpret the past.

ACKNOWLEDGMENTS

This paper was originally presented in a symposium entitled "Applications of Mammalian Mortality Patterns for the Study of Prehistoric Human Foraging Ecology" at the 54th Annual Meeting of the Society for American Archaeology in Atlanta, Georgia, April 5-9, 1989. I thank Mary Stiner for her invitation to participate in the symposium and for her efforts in seeing the manuscript through to publication. I would also like to acknowledge the generous support of the School of American Research in Santa Fe, New Mexico; the paper took on its final form there while I was on sabbatical leave as an SAR Resident Scholar.

4

Examining and Refining the Quadratic Crown Height Method of Age Estimation

Diane Gifford-Gonzalez

INTRODUCTION

As chapters in this volume attest, archaeologists have increasingly turned to analyses of mortality data from archaeofaunas. Their goals have included evaluating the role of hominids in creating bone accumulations and elucidating ancient foraging behavior. Such research stems in part from ecological studies of mammalian mortality (e.g., Caughley 1966; Gavin *et al.* 1984; Peterson *et al.* 1984; Sinclair 1977) and augments a long-standing archaeological tradition of studying mortality patterns of domestic animals from later prehistoric sites (Ewbank *et al.* 1964; Ducos 1968; Grant 1982; Payne 1973). Major publications on ungulate mortality profiles in hunter-gatherer sites include those of Frison, Reher, and associates for North American plains hunters, (Frison 1978b; Frison and Reher 1970; Frison and Todd 1987; Todd and Hofman 1987), Spiess (1979) for Old and New World caribou hunters, and Klein and his co-workers (Klein 1978, 1981b, 1982a, 1982b; Klein *et al.* 1981, 1983; Klein and Cruz-Uribe 1983a) for European and African paleolithic hominids.

All analyses of mortality patterns depend on reasonably accurate estimations of age at death, and the precision required increases with the number of age cohorts considered. This chapter evaluates one such method, age estimates based on enamel crown heights measurements, using known-age bison samples. Although wildlife biologists have experimented with the enamel crown height method of aging dentitions for forty years (Severinghaus 1950; Lowe 1967; Spinage 1971, 1972, 1976), it has received attention among archaeologists largely through the publications of Richard Klein and his associates. They have used the method as a basis for reconstructing mortality profiles for hypsodont (high-crowned) hoofed animal species, including European red deer and various southern African bovids and equids, recovered from archaeological sites. Klein and Cruz-Uribe (1984) present computer programs for calculating age-at-death from crown height measurements and constructing mortality profiles from these statistics.

Here I report results of an independent evaluation of this method using two samples of known-age bison teeth. I undertook the evaluation because I had encountered problems applying the aging method in a study of Neolithic cattle teeth from East Africa. My results indicate that the formulae published by Klein *et al.* produce problematic estimates for bovine teeth, which should be noted by researchers who contemplate using this approach. This study should not, however, be read as a simple cautionary tale, but rather as an attempt to evaluate and refine fine-grained aging methods based on enamel crown heights. Problems with the method can be isolated and alternative approaches proposed; this chapter reports on a first step in that process.

I begin with an overview of the quadratic method and some of the potential advantages of aging methods that use remnant crown height. Discussion of problems encountered when applying the quadratic method to archaeological cattle specimens follows. Next, results of the known-age bison study are presented. The concluding section discusses implications of the study for constructing and interpreting mortality profiles involving many age cohorts.

THE QUADRATIC METHOD

The crown height age estimation method developed by Klein calculates age at death through a quadratic equation describing a curvilinear relationship between actual age and attrition of the tooth crown (Klein, *et al.* 1983; Klein and Cruz-Uribe 1983a). Two formulae for calculating age-at-death from enamel crown height are used: one for deciduous teeth and one for permanent teeth. The formula for deciduous teeth is

$$AGE=AGEs-2AGEs(CH/CHo)+AGEs(CH^2/CHo^2)$$

The formula for permanent teeth is

$$AGE = AGEpel - 2(AGEpel - AGEe)(CH/CHo) + (AGEpel - AGEe)(CH^2/CHo^2)$$

where

> *AGEs* is the age at which the deciduous tooth is shed;
> *AGEe* is the age at which the permanent tooth erupts;
> *AGEpel* is the potential ecological longevity (maximum lifespan) of any member of the species;
> *CH* is the height of the crown of the tooth at the time of death; and,
> *CHo* is the crown height of the crown when it is fully formed but unworn.

Both formulae describe parabolas approximating variation in rates of dental attrition over the lifespan of the animal.

Klein notes that the quadratic equation is actually an algebraic transformation of a formula devised by Spinage (1971), which states that the height of a hypsodont tooth in wear is reduced as a square root function of its preceding height. The Spinage model predicts that teeth initially will wear swiftly, but at ever-slower rates in later life.

Klein and Cruz-Uribe (1984) outline the advantages of age estimation methods based on remnant crown-height, the most pertinent of which are summarized here:

1) Unlike sectioning of teeth for counting cementum or dentin annuli (viz. Erickson and Seliger 1969; Mitchell 1963; Spinage 1973; Spiess 1978), measuring remnant crown height is nondestructive and takes relatively little time to accomplish.
2) Unlike the annulus-based method, the crown height method can be applied to all moderately high-crowned teeth, regardless of whether they have been diagenecally altered through fossilization.
3) Being metrical, the method is less subjective than assessing occlusal wear patterns, another commonly-used method of estimating age-at-death (e.g., Ewbank *et al.* 1964; Grant 1982; Payne 1973; Deniz and Payne 1983).
4) The metrical data obtained by the crown height method are more readily manipulated than are those from occlusal wear analysis.

Klein's Check of the Method Using Known-Age Wapiti Dentitions

In developing and testing the quadratic method of age estimation, Klein *et al.* (1983) undertook a check using a known-age sample of 170 North American wapiti (*Cervus elaphus canadensis*) from Yellowstone National Park; the wapiti is the North American counterpart of the European red deer. Measurements of crown heights were taken for mandibular cheekteeth (including deciduous lower premolars), and ages were estimated and compared against each individual's documented age at death. The investigators drew two main conclusions from their analysis:

1) Errors in age estimates based on wear were sometimes substantial, but they tended to fall more or less evenly around the true ages of the teeth evaluated.
2) The quadratic formula did not yield equally precise estimates for all portions of the life course.

Both points merit more detailed examination. First, although individual variation in crown heights of any one-year age class was sufficiently great to produce substantial errors in age estimates, variation fell relatively evenly around the actual ages. In referring to the errors in age estimates that such biologically based variability would engender, Klein and Cruz-Uribe (1983a:76) note that the method appeared "to underestimate the true age about as often as they overestimate it." They go on to infer that individual age estimation errors "will tend to cancel each other out when the ages are grouped into the relatively broad age classes we use." In sum, they noted no *directional* biases in the errors of age estimates derived from their formulae.

Klein and Cruz-Uribe's "broad age classes" are in 10%-of-lifespan intervals, rather than in years. In their 1981 paper, Klein *et al.* concluded that, given the potential ecological longevity of wapiti in the wild, errors in age estimates of plus or minus 10 percent of that span (i.e., ± 19.2 months) could be expected and would not undermine the utility of the method. The authors stress that the *overall* shape of the mortality profiles, rather than the year values of the points contained therein, is most significant (Klein *et al.* 1983).

The second major finding of the check using known-age wapiti was that the quadratic equation did not produce satisfactory age determinations for all ages in the total lifespan. Klein *et al.* (1981) state that the method they were using at that time was most likely to grossly mispredict ages based on both recently erupted and heavily worn teeth. They urged caution in working with dental samples containing many teeth in these two categories (Klein *et al.* 1981:30). In a later publication, Klein and Cruz-Uribe (1984) explain that, among older animals in the known-age wapiti sample, posterior teeth tend to underestimate real age (especially third molars), whereas first molars tend to overestimate it.

It follows from these observations that the *modal age* of specimens in any given sample may influence the relative "success" of the quadratic formula in aging them. In the known-age wapiti sample, 90% (153 of 170) of the individuals were six and half years of age or younger when they died. These individuals therefore fall into the first 40% of the maximum lifespan, the span over which the quadratic formula normally describes a near-linear relationship between dental attrition and age (cf. Klein *et al.* 1983). Dental samples containing more specimens representing the extreme ends of an age distribution might be expected to yield different, less reliable, results. Klein *et al.* do not specify precisely how much of the potential lifespan is affected by these types of errors, nor do they discuss the nature and causes of the potential mispredictions in detail.

This chapter is mainly about errors at the ends of age distributions. My results with bison and cattle dentitions indicate that the Klein quadratic formula indeed does not produce consistently acceptable age estimates at the earlier end of the age range. Substantial mispredictions occur for the first three to four years of wear of any given molar. This means that the reliability of age estimates is of concern not only for dental samples containing many very young animals, but also for those including adults in the younger part of the "prime age" group.

THE NEOLITHIC CATTLE STUDY

I began using the crown height age estimation method in the context of studying Neolithic-age cattle teeth from Kenya in the early 1980s, ultimately with the idea of also applying the method to sheep, goats and wild ungulate remains. I chose the method because I was looking for a swift, nondestructive method for aging the thousands of isolated teeth recovered from these sites. Teeth of larger ungulates, such as cattle, wildebeest and zebra, nearly always occur as isolated specimens in Kenyan Neolithic sites, probably due to heavy culinary processing, post-discard trampling and geologically-caused reduction of bone. This situation contrasts with the frequent preservation of mandibular or maxillary dental arcades in some European sites containing domesticated animal remains, and in some bison mass kill sites from North America. In less fragmented assemblages, aging by overall eruption and occlusal wear patterns of more complete dental arcades is commonly used with considerable accuracy (Ewbank *et al*. 1964; Grant 1982; Frison and Reher 1970; Frison 1978b; Frison and Todd 1987). Other taphonomic factors also reduce opportunities for aging of Neolithic teeth by sectioning methods: many measurable molar crowns have lost their roots bearing the cementum deposits needed for performing annuli counts.

An Internal Check of Neolithic Cattle Age Estimates

When I began working with the Neolithic cattle dentitions, I thought it best to check the quadratic method's consistency in aging this species. Cattle have a different eruption sequence and a far greater degree of hypsodonty than wapiti, which might affect their patterns of dental attrition. Unlike the situation for cervids, the M3 of cattle and other members of the bovid tribe erupts *before* the deciduous fourth premolars are shed and replaced by permanent premolars (Silver 1969). Bovines (including cattle) have high-crowned molars relative to those of wapiti. This can have important consequences for the pattern of eruption and wear in the respective taxa. Moreover, wapiti molar crowns are fully erupted earlier in the lifespan of an animal. Molars of cattle, on the other hand, continue to erupt from their crypts over a longer segment of the animals' lifetime, in pace with attrition on the exposed section of the crown. These developmental differences persuaded me that biologically-based disparities in patterns of tooth eruption and wear might exist between bovids and cervids.

I was unable to locate an adequate sample of known-age cattle dentitions with which to check the method. I opted to take another approach that entailed comparing age determinations based on several teeth from one individual's maxillary or mandibular dental arcade ("tooth row") in the archaeological sample. This would not yield any "real" age referents, but would allow me to assess the internal consistency of age determinations from different teeth from the same animal. While complete tooth rows were relatively scarce in the archaeological assemblages studied, I was able to assemble 70 partial dental arcades from seven Neolithic sites. Each dental arcade contained two or three molars.

Because I lacked external age referents, I asked whether ages estimates derived from different teeth in the same jaw fell into similar age ranges; specifically, are age estimates for second and third molars more than plus or minus 10% of potential ecological longevity of the first molar's age determination, as this is the range of error accepted by Klein *et al.* (e.g., 1983; Klein and Cruz-Uribe 1984)? In other words, did the second and third molar age estimates fall into adjacent 10% age classes to the first molar age, or did they fall farther away from it?

I found that upper tooth rows in the cattle sample were modally at earlier stages of eruption and wear than the lower tooth rows, probably due to vagaries of excavation sampling. For the upper molars, about 85% of the sample (based on the first molar age estimates) fell into the same first 40% of maximum longevity, a pattern analogous to Klein *et al.*'s wapiti sample. In contrast, only 55% of lower molars fell into the first 40% of the maximum lifespan. This disparity highlighted and further clarified some of the problems with age estimates at both ends of the age spectrum.

When I compared the age estimates for M1 to those for M2 and M3, some interesting patterns emerged. Figure 4.1 displays age estimates derived from molars of the same maxillae and mandibles. With the estimated potential ecological longevity for African cattle set at 144 months (cf. Dahl and Hjort 1976), differences of 30 months or more in age estimates would exceed the "plus-or-minus 10% of lifespan" acceptance range stipulated by Klein *et al*. As would be expected, ages derived from molars of more worn dentitions (M1 aged at greater than 48 months) frequently differed considerably, with the greatest disparities in age estimates occurring for the most worn dentitions. Unexpected, and more problematic, were the consistent patterns of difference in age among the teeth of younger animals in the upper tooth row sample. Maxillary second molars in very early wear (eruption occurs around 18 months) rested alongside first molars that aged to slightly over six months (see Fig. 4.1). Moreover, of the maxillae with third molars in very early wear, adjacent anterior molars aged close to their respective ages of eruption (6+ and 18+ months, respectively). In all cases, the posterior molar was fully formed and in early wear, indicating that the animal was slightly past the average age of eruption for the molar (in the case of M3's, around 24 months). Nothing in the patterns of occlusal wear appeared to distinguish these "under-aged" anterior molars from counterparts in animals known to be 6 or 18 months of age. The mandibular molar sample had only tooth row in this age range, but it displays a similar age estimate pattern (Fig. 4.1, arrow).

Given these observations, I hypothesized that attrition on the first and second molars of cattle is slow between their respective times of eruption relative to that of the third molar. This hypothesis made sense from a functional perspective, because of the relationship between deciduous

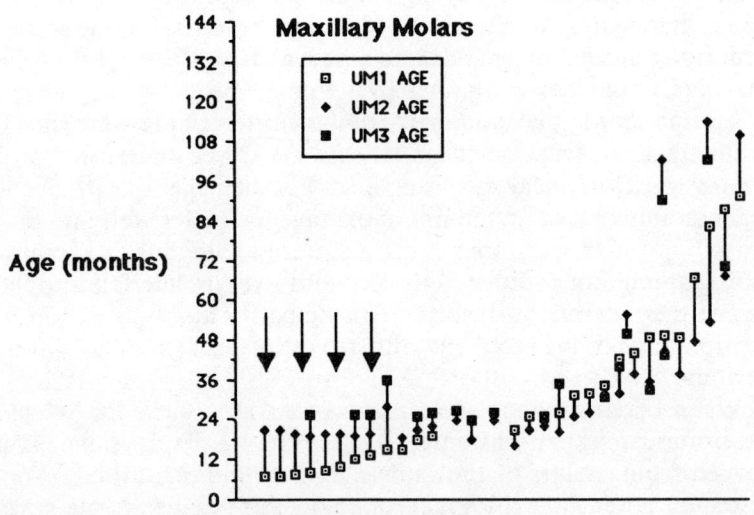

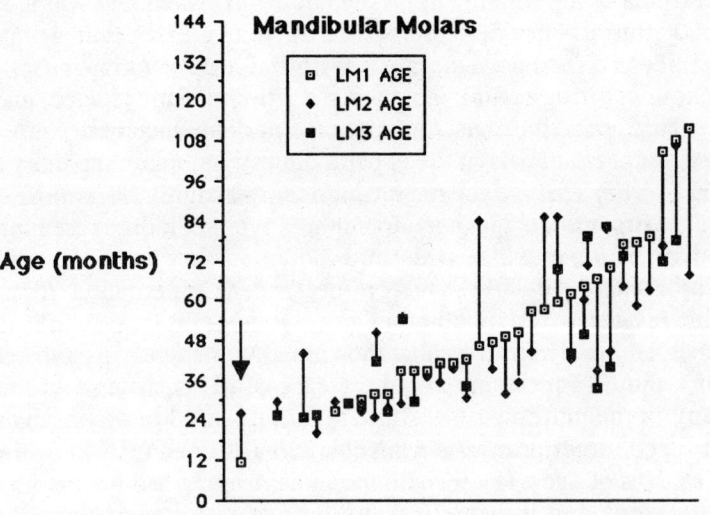

Figure 4.1. Differences in quadratic-derived age estimations for molars in Neolithic cattle partial dental arcades. Lines connect determinations for molars of same jaw. Arrows indicate age range discussed in text.

premolar shedding and molar eruption. In cattle, for example, the molariform deciduous P4's are not shed until about a year *after* the M3 erupts. From an evolutionary perspective, one might argue that selection favors individuals that maximize the use-life of their molars by sparing them from heavy attrition while dP3's and dP4's were still functional, and only putting them into heavy wear after the deciduous teeth were shed.

One might object that, even if this hypothetical molar-sparing phenomenon caused substantial under-aging in younger cattle, such eruption-related differences in age determination are not significant for an analytical point of view. Obviously, this depends in part on how finely one chooses to subdivide the idealized lifespan. In the case of the offset in eruption age between M1 and M2, this difference is about 12 months, between M1 and M3, about 18 months. Building on Klein *et al.*'s discussion of acceptable ranges of error in age estimation, one might argue that both fall well within the 10% error zone for cattle's potential ecological longevity. Thus, although it is troubling to have an 18-month-old bovine's M1 indicating a age of only about 7 months, this difference could be said to fall within acceptable limits. It could also be argued that, since the *shape* of the curves derived from any one of these three molars should theoretically be similar, these systematic differences will have little impact on the interpretation of mortality curves based on different molars.

Yet, these differences in age estimates raise some troubling questions about how to interpret and use the many *isolated* first, second, and third molars displaying only light wear in archaeological assemblages. Regardless of whether analysts use crown heights or occlusal wear to age individual teeth, they might assume that the three different molars, each of which was only slightly worn, *necessarily* represent different individual animals and try to work with the data as such. Inferences about hominid behavior using *multiple-cohort* mortality profiles constructed from a cluster of newly erupted M1's (animals about six months old) would probably differ substantially from those based on a set of recently-erupted M2's (animals about 18 months old) or M3's (animals about two years old).

Some Kenyan Neolithic cattle dental samples in fact produced modes in age estimates approximating the patterns from associated molars of cattle slaughtered just as the M3 was coming into wear. Figure 4.2 shows a set of six mortality profiles from the GvJm44 site, at Lukenya Hill, south of Nairobi. Figure 4.3 shows three profiles from the site of Narosura, on the edge of the Loita Plains in southwestern Kenya. Each profile was constructed using the quadratic formula from different molars, combined with ages from the most common *deciduous* premolar in the same assemblage (a practice outlined by Klein and Cruz-Uribe 1984). For

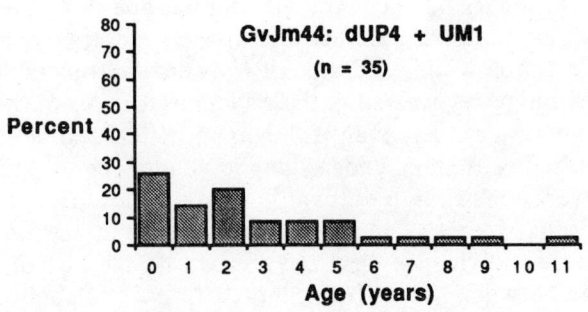

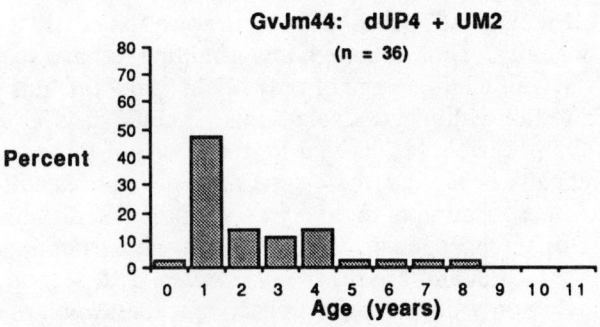

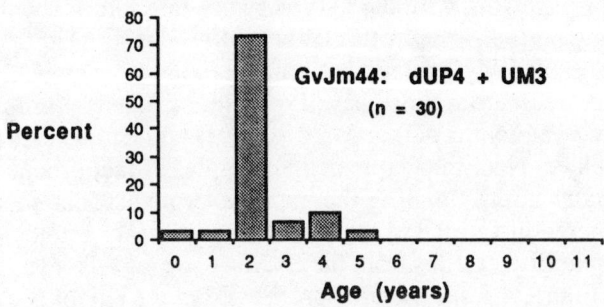

Figure 4.2a. Mortality profiles for Neolithic cattle teeth from GvJm44, Lukenya Hill, Kenya: maxillary molars.

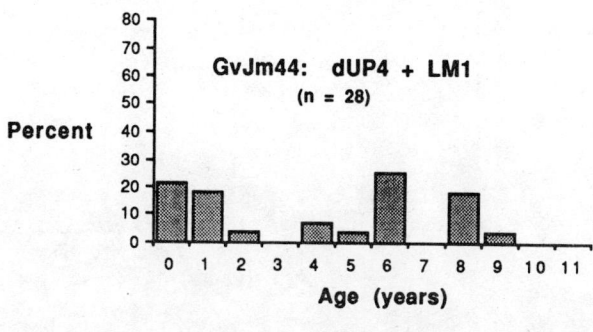
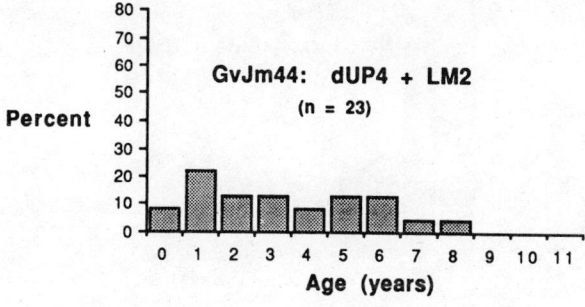
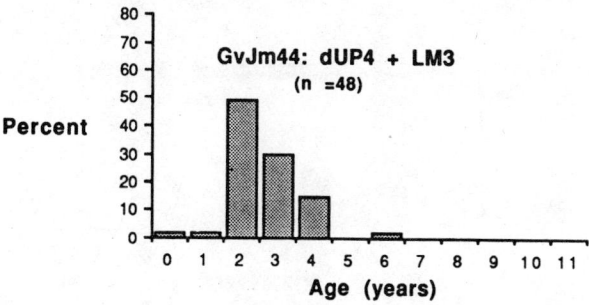

Figure 4.2b. Mortality profiles for Neolithic cattle teeth from GvJm44, Lukenya Hill, Kenya: mandibular molars.

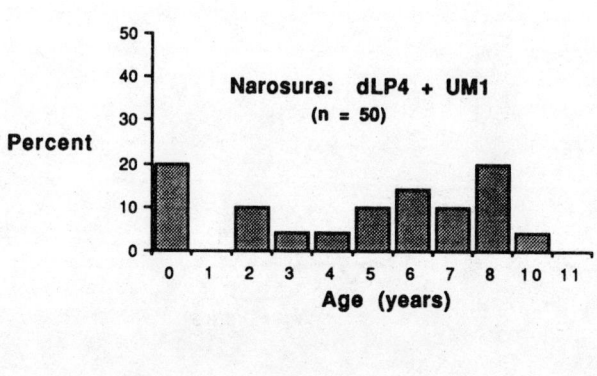

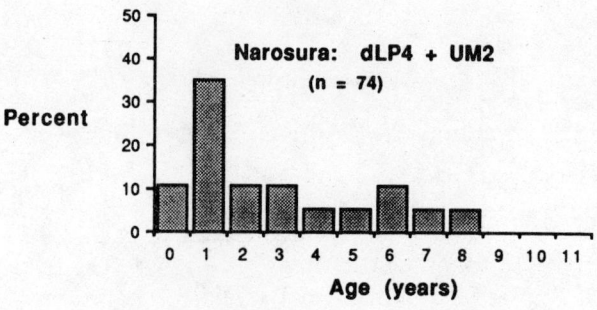

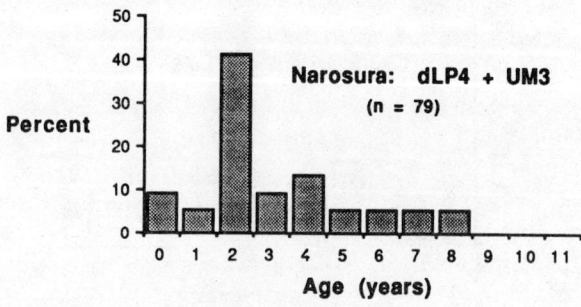

Figure 4.3. Mortality profiles for Neolithic cattle teeth from Narosura, Loita Plains, Kenya, upper molars only.

upper molars, one can see peaks in age-at-death around the eruption periods for both M2 and M3. In light the findings from aging the partial dental arcades, I suspected these patterns could result from aging isolated teeth that actually came from the *same* mandibles, in animals killed around 24 months of age.

Are these distributions different enough to worry about which one is the "right" one? For 10 or more age cohorts, statistical assessments indicate that the answer was "yes". When the upper molars from GvJm44 are compared using the Kolmogorov-Smirnov test (following Klein and Cruz-Uribe 1984), M2 and M3 are significantly different ($p<0.01$). Among the mandibular molars from GvJm44, M1 and M3 are different at more significant level ($p<0.001$). However, mortality profiles for any given upper molar and its lower mate are not significantly different. All three upper molar profiles from the larger Narosura sample are significantly different.

In light of the results of the tooth row checks, I suspected that the statistically significant differences in clusters of age-at-death represented an interaction of biologically-determined variability in molar wear and the quadratic formula. I also wondered what the useful limits of the crown height-based quadratic method for estimating age would be, even in corrected form.

THE KNOWN-AGE BISON STUDY

Known-age cattle dentition samples representing more than the first 24 months of life either were not available for comparison or practical for study in the context of this research, but I was able to use known-age dental samples from two bison (*Bison bison*) populations in the Department of Zoology and Physiology, University of Wyoming.[1] Bison are members of the same bovid tribe as cattle, they are interfertile and have similar feeding habits. Bison and cattle teeth are difficult to distinguish morphologically and, although bison have slightly later eruption times for M3 than cattle, I assume here that the processes of eruption and wear are very similar between the two species. One bison sample is from the Yellowstone National Park herd, which subsists exclusively on wild forage. Animals in the Yellowstone herd have been regularly tagged, aged and monitored for over two decades. Cows regularly reproduce into their early 20's, and bulls survive at least as long (M. Meagher, personal communication, see also Meagher 1973).

The other sample is from the free-ranging herd at "Bison Pete" Gardner's ranch in Wheatland, southeastern Wyoming. The bison are fed alfalfa cake in the winter months, but forage on their own over rest of the year. Unlike the situation for most cattle, older bison in the Wheatland herd are culled only due to prolonged illness or senility. As with the Yellowstone group, Wheatland bison cows may continue to calve into their early 20's, and some bulls live 25 years or more (D. Walker, personal communication).

Bison reproductive patterns can be expected to exert a strong influence on the form of any age-indexed data. Most bison calve within two weeks of one another in spring; in the northern plains of the United States, this occurs in late April through early May (Frison and Reher 1970; Todd and Hofman 1987). Age cohorts are thus tightly defined. Most of the animals in these samples are autumn culls, the youngest age class being 6 months old, and each succeeding age class corresponding to yearly increments therefrom. The two death samples have different age structures, as shown in Table 4.1.

Measurements are for left mandibles only, except where the left side was absent or damaged and then the right mandible was substituted. Each mandible was sectioned from the lingual side of the dentary to expose the junctions of molar enamel crowns and roots. Measurements were taken on all three molars. Crown height and basal enamel breadth were measured to the nearest tenth of a millimeter. Crown height measurements were taken on the same points used by Spinage (1971:210): from top to base of the enamel in the "saddle" between the first and second lobes of the teeth on the lingual side of the tooth. This puts the measurement between the metaconid and paraconid "peaks," and therefore differs from the measurement sites used by Klein (cf. Klein *et al.* 1983), and by Todd, Hofman and others (cf. Frison and Todd 1987). I found the Spinage technique more useful because it could be applied to a greater number of damaged teeth in the archaeological samples I worked with. I found that all three measures are essentially interchangeable in terms of derived age estimates for Neolithic cattle.

One Sample or Two?

For the purposes of this study, I have kept the Yellowstone and Wheatland dental samples separate. Assessments of differences in molar size and rates of wear between the Yellowstone and Wheatland samples have yielded equivocal results and counsel caution in their use as a pooled

Table 4.1. Actual age distributions of bison in the Wheatland and Yellowstone dental samples.

age in months	Wheatland	Yellowstone
6	1	15
30	14	2
42	18	7
54	15	11
120	0	4
Total	65	52

sample. Of greatest concern were inter-population differences in molar size and rates of dental attrition. Major differences in molar size between the two populations could contribute to greater ranges of age-specific crown heights in a pooled sample and blur age-specific attritional trends in either. Variation in wear rates between the two populations would also obscure age-specific wear trends in either. Some dental samples for prehistoric bison in other regions suggest varying rates of molar wear (Speth 1983; Todd and Hofman 1987), and similar variation in rates of wear have been documented by Deniz and Payne (1982) among populations of Angora goats. Inter-population disparities in wear rates are ascribed to differing amounts of abrasive matter in the diet, both in the form of dust and in plant tissues.

The question of tooth size was addressed by comparing basal enamel breadths for first, second, and third molars in the two samples, using a two-tailed t-test (Table 4.2). Basal crown breadth is not appreciably affected by either occlusal or interstitial wear over the lifespan. Results for the first molar refute the null hypothesis that no statistically significant difference exists between the two samples, whereas t-test results for the second and third molars support it. Since the first molars were drawn from the same mandibles as second and third molars, this was an unanticipated mix of results. An attempt to assess the possibility of disparate wear patterns also led me to keep the sample separate.

In fact, differential rates of dental wear could not be directly assessed with the data at hand. Each age cohort represents individuals sampled at a given point in their respective tooth attrition histories, and significant dif-

Table 4.2. Two-sample *t*-test of molar breadth for Wheatland and Yellowstone samples.

molar	mean	SD	N	*t*-statistic	df	*p*
Wheatland M1	21.36	1.38	65			
Yellowstone M1	19.87	1.42	51	-5.699	114	<.001
Wheatland M2	26.92	1.67	64			
Yellowstone M2	26.78	2.14	37	-0.365	99	>.50, <.90
Wheatland M3	39.15	3.12	48			
Yellowstone M3	39.78	2.82	25	0.845	71	>.50, <.90

ferences in the mean crown heights of molars in comparable age cohorts can only suggest the operation of some biasing process — either sampling error or differential attrition. Nonetheless, significant inter-population differences in age-specific crown heights at more advanced states of wear were considered grounds for studying the two populations separately, whatever the cause of that difference. Tables 4.3a, b and c present two-tailed *t*-test results for molar crown heights of each comparable age cohort in the Yellowstone and Wheatland samples. Crown heights in the Yellowstone and Wheatland cohorts in early wear (e.g., 18 months) are similar, whereas the Wheatland sample consistently displays lower crown heights among older cohorts. While it is tempting to interpret increasing levels of significance in these statistics as the effects of cumulatively swifter rates of molar wear in the Wheatland herd, the data cannot establish this.

Evaluating the Quadratic Method with the Known-Age Samples

How does the quadratic method of age estimation work on the known age bison samples? Does the molar-sparing pattern of early wear seen in Neolithic African cattle also apply to bison, and how might this affect quadratic-derived age estimates? Before addressing these questions, some variables called for by the equation must be explained in more detail.

I have used 25 years (300 months) as the potential ecological longevity ($AGEpel$) of bison. Although some captive bison have survived to over 40 years (Jennings and Hebbring 1983), 25 years is deemed a realistic

Table 4.3a. Two-sample *t*-test of M1 crown heights, by age class, for Wheatland and Yellowstone samples.

sample	mean	SD	N	*t*-statistic	df	*p*
18-19 MONTHS:						
Wheatland	39.7	2.8	17			
Yellowstone	38.6	2.0	12	-1.168	27	>0.2
30 MONTHS:						
Wheatland	34.8	2.7	14			
Yellowstone	33.1	2.4	2	-0.839	14	~0.4
42 MONTHS:						
Wheatland	32.8	3.4	18			
Yellowstone	35.9	2.3	7	2.209	23	0.05
54 MONTHS:						
Wheatland	18.5	5.2	15			
Yellowstone	25.2	6.1	11	3.018	24	<0.01

Table 4.3b. Two-sample t-test of M2 crown heights, by age class, for Wheatland and Yellowstone samples.

sample	mean	SD	N	*t*-statistic	df	*p*
18-19 MONTHS:						
Wheatland	55.0	2.0	17			
Yellowstone	54.6	1.9	12	-0.541	27	>0.5
30 MONTHS:						
Wheatland	52.0	2.3	14			
Yellowstone	49.7	.42	2	-1.375	14	~0.5
42 MONTHS:						
Wheatland	50.0	3.8	18			
Yellowstone	51.7	3.2	7	0.793	23	~0.5

(continued)

Table 4.3b. (Continued)

sample	mean	SD	N	t-statistic	df	p
54 MONTHS:						
Wheatland	30.6	7.7	15			
Yellowstone	40.9	7.7	11	3.370	24	<0.01

Table 4.3c. Two-sample t-test of M3 crown heights, by age class, for Wheatland and Yellowstone samples.

sample	mean	SD	N	t-statistic	df	p
30 MONTHS:						
Wheatland	53.1	2.4	14			
Yellowstone	52.1	0.1	2	-0.572	14	<0.9, >0.5
42 MONTHS:						
Wheatland	53.7	1.4	18			
Yellowstone	53.5	2.1	7	-0.279	23	<0.9, >0.5
54 MONTHS:						
Wheatland	41.1	6.9	15			
Yellowstone	46.6	8.0	7	1.658	20	<0.2, >0.1

estimate by two persons who have worked with modern bison populations, including the Wheatland and Yellowstone herds (M. Meagher, D. Walker, personal communications). I have used 6, 18, and 30 months as ages of eruption ($AGEe$) for first, second, and third molars (Frison and Reher 1970; Todd and Hofman 1987).

Crown heights at eruption (CHo, fully formed but unworn) for each molar were calculated by taking the mean heights of unworn molars for each bison sample, except for the 6-month sample because Wheatland has only one M1 at eruption. The data on M1's for Wheatland and Yellowstone were combined, as this seemed more sensible. This potentially presents a methodological problem, because the M1's from these two samples display significant levels of difference in mean breadths; one could argue that crown heights at eruption could also be different. However, I have found no significant correlation between basal breadth and crown height in either of these samples, and therefore no logical basis for avoiding use of a pooled mean CHo for Wheatland M1's.

With these variable values, I used the quadratic formula to calculate ages for each molar. I then expressed the difference between the age estimate derived from the quadratic formula and the actual age as "percent error" in the following way:

$$\% \ Error = (Estimated \ Age - Actual \ Age)/Actual \ Age$$

An age determination that underestimates the actual age of the animal will be expressed as a negative percent of actual age, and an overestimate will be expressed as a positive percentage.

Figures 4.4 and 4.5 present percent error statistics for quadratic formula age estimates on M1's at the different actual ages represented by the sample. I do not provide the estimates for the 6-month age class here, because these clustered closely around the 6-month value for age-at-eruption. Deviations in estimates at this age reflect the inherent variation in crown height at eruption in the samples in relation to the mean CHo statistic. At 18 through 42 months, both the Wheatland and Yellowstone M1 age estimates show a strong tendency to *underestimate* age. This bias continues until 54 months of age, at which point the estimates derived from the quadratic formula show greater dispersal around the true age and a tendency toward *overestimation* of age. In other words, between 42 and 54 months, dental attrition intensifies, producing in some animals much higher levels of wear than predicted by Spinage's model.

Figures 4.6 and 4.7 show age estimates for M2's from the Wheatland and Yellowstone samples, again with dentitions at the age of eruption excluded. The tendency toward underestimated ages is strong until around 54 months, where case estimates become more dispersed, with a tendency toward overestimation. Distributions of age estimates for M3's show generally similar tendencies (Figs. 4.8 and 4.9). However, the shorter spans of wear represented in the M3 samples, especially Wheatland, probably do not allow the underestimate-to-overestimate transition to be fully expressed.

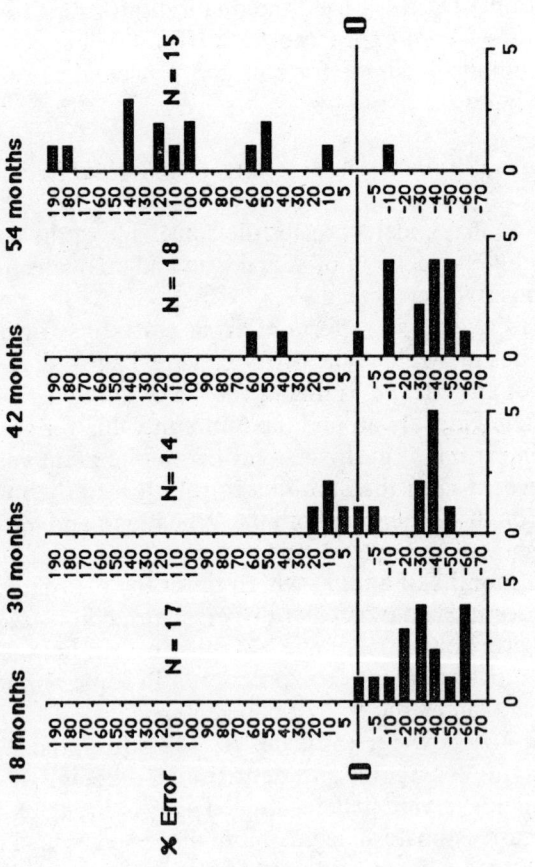

Figure 4.4. Percent error of quadratic-derived age estimates for M1, Wheatland sample, grouped by actual ages.

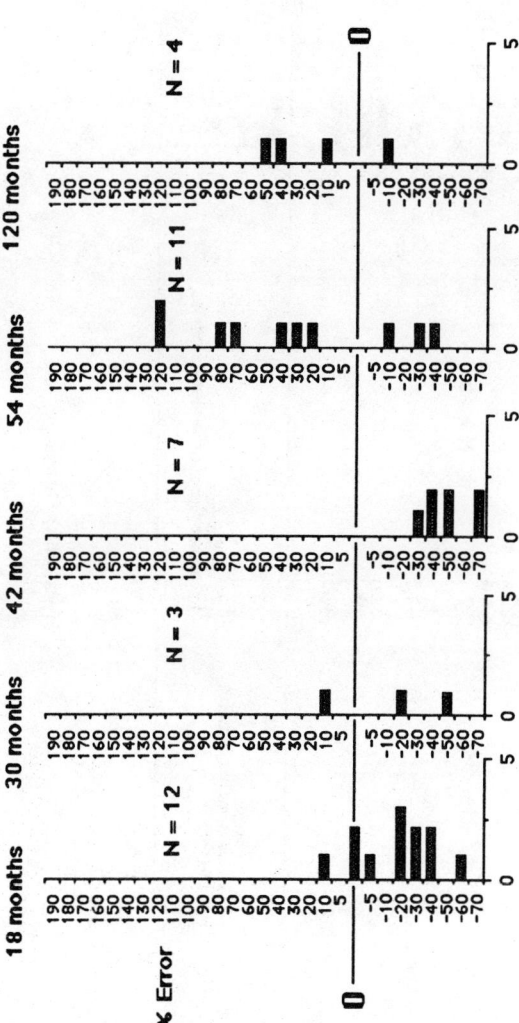

Figure 4.5. Percent error of quadratic-derived age estimates for M1, Yellowstone sample, grouped by actual ages. Age at eruption excluded (these are, by definition, nearly 100% accurate).

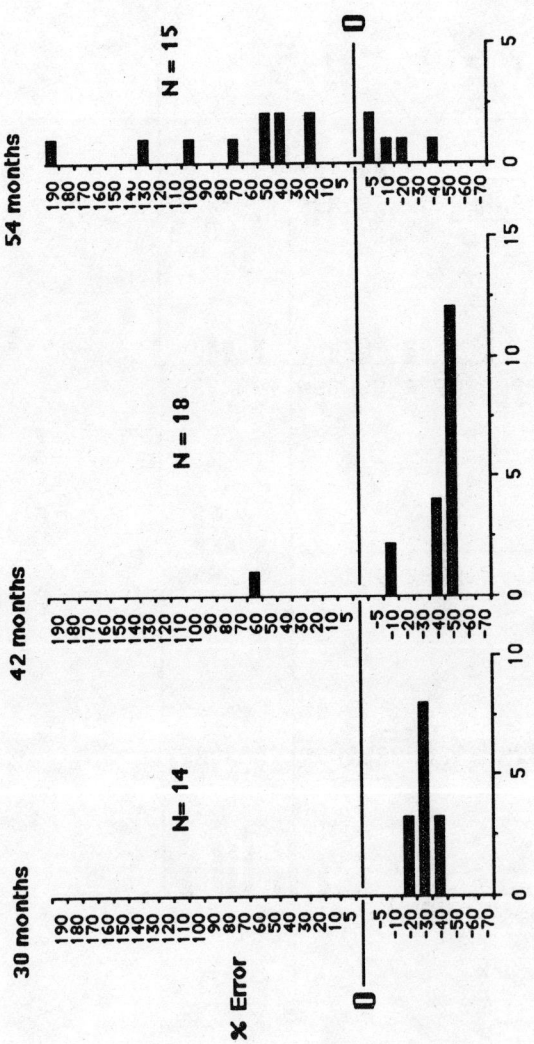

Figure 4.6. Percent error of quadratic-derived age estimates for M2, Wheatland sample, grouped by actual ages. Age at eruption excluded (these are, by definition, nearly 1000% accurate).

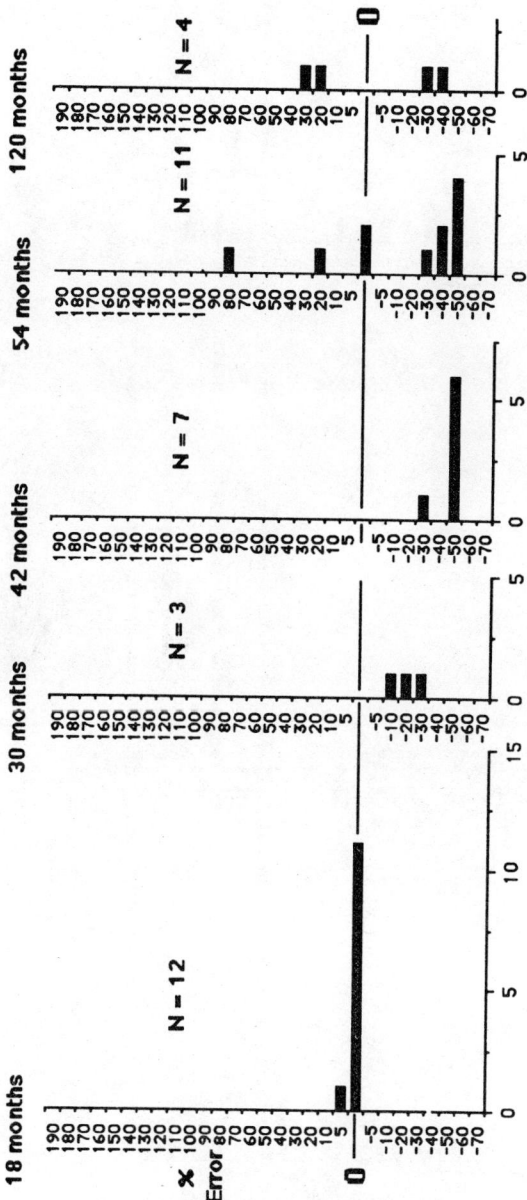

Figure 4.7. Percent error of quadratic-derived age estimates for M2, Yellowstone sample, grouped by actual ages. Ages at eruption excluded (these are, by definition, nearly 1000% accurate).

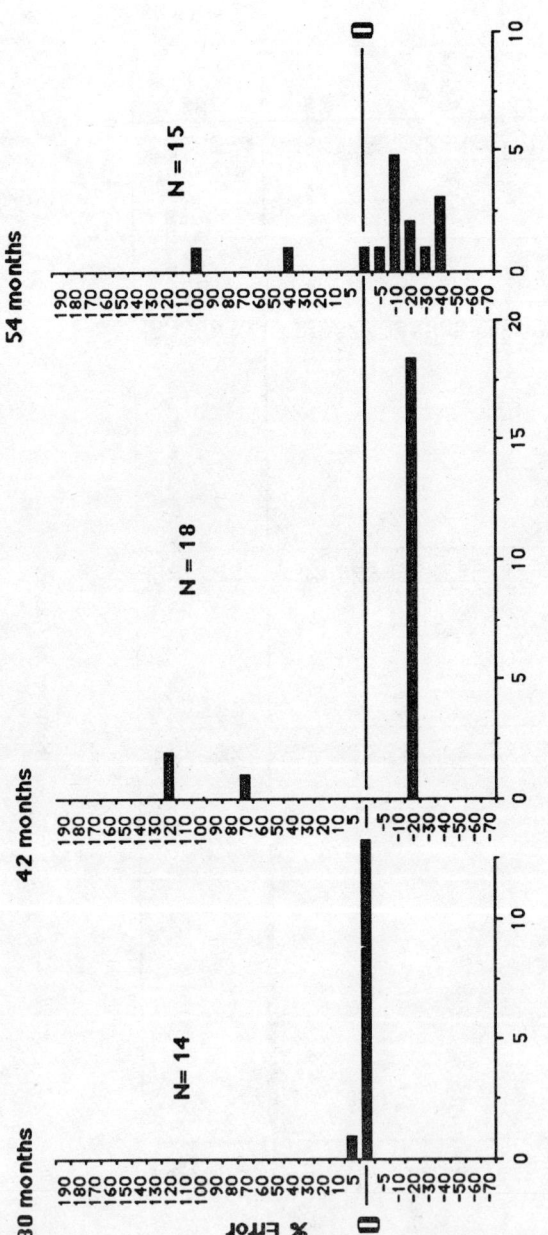

Figure 4.8. Percent error of quadratic-derived age estimates for M3, Wheatland sample, grouped by actual ages.

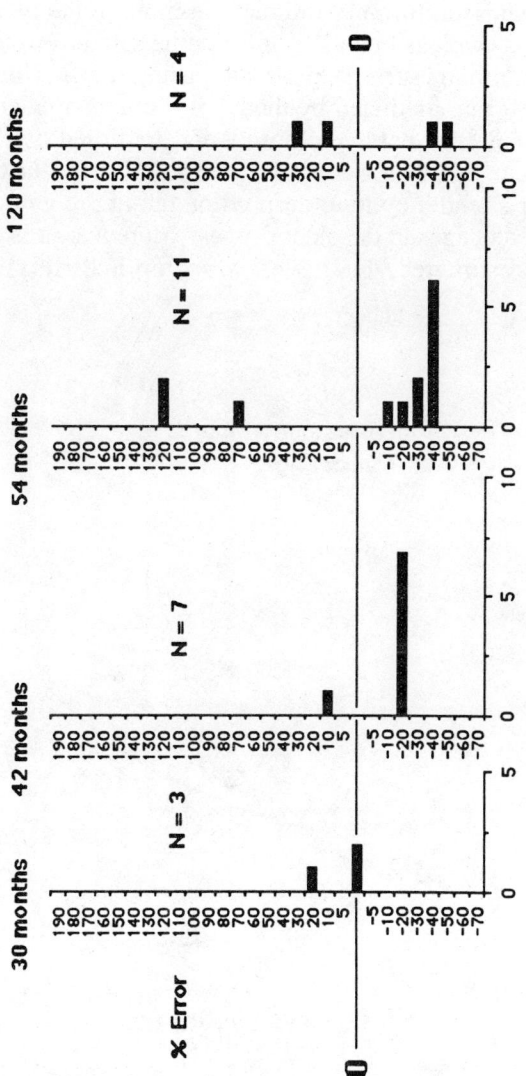

Figure 4.9. Percent error of quadratic-derived age estimates for M3, Yellowstone sample, grouped by actual ages.

Molar-Sparing and Early Dental Attrition

These striking and consistent patterns in percent error distributions are caused by a fundamental difference between the pattern and rate of wear predicted by the quadratic formula and that observed in the two known-age bison samples, as well as in the Neolithic cattle samples from Kenya. Specifically, bison molars wear *very slowly* during the first few years, rather than the swift rate predicted by the Klein-Spinage model. Figure 4.10 shows the difference between tooth wear predicted by the Klein model and the actual pattern for the Yellowstone M1's; it illustrates the quadratic formula's tendency to underpredict age in the youngest age classes and overpredict ages in the older classes. Moreover, it is clear that ages will be underestimated, due to very slow tooth attrition up to 42 months of age.

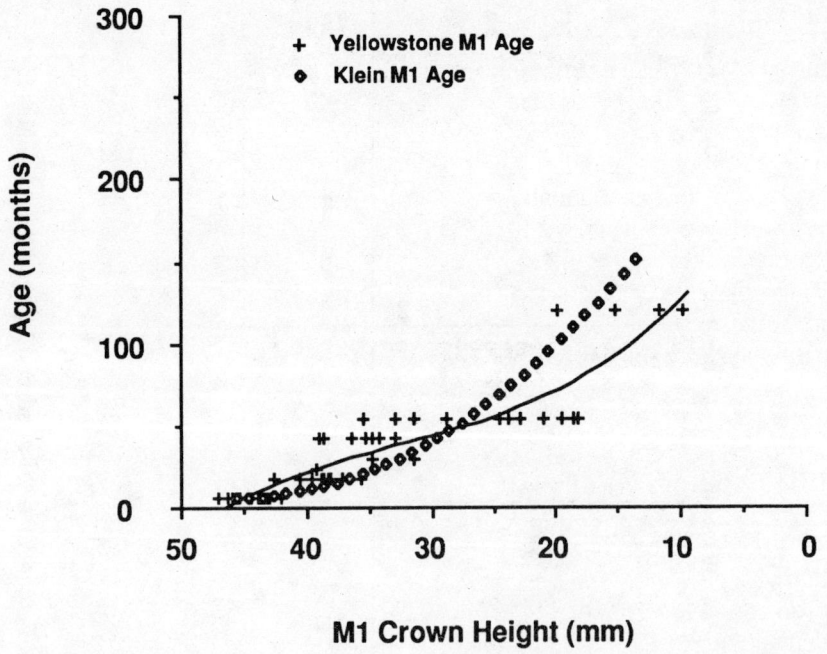

Figure 4.10. Curvilinear regression of Yellowstone M1 data compared to theoretical model described by quadratic formula, showing difference in predicted rates of attrition early in wear.

The analysis of the known-age bison samples supports the molar-sparing early attrition pattern hypothesized for bovines in the Neolithic cattle study. It will be remembered that Klein and his co-workers warned against relying on their method for the very early span of molar wear. The bison data indicate that the danger of mistaking anterior teeth for those of calendrically-younger (and physically less mature) individuals exists through the fourth or fifth year of these large ungulates. This problem is especially vexing with isolated teeth.

Molar-saving attrition may be more extreme in hyper-hypsodont species for adaptive reasons. Hypsodonty itself is seen as an adaptive response to high proportions of abrasive material in the diet. However, it could either result from simple allometric effects, in which higher-crowned teeth display more extreme sparing tendencies, or it could result from selection operating specifically on tooth development and eruption, which similarly optimizes the amount of grinding surface available to the animals over time.

A possible objection to my findings on the molar-sparing phenomenon relates to the site chosen for tooth measurement. Recall that the bison molars were measured in the saddle area between the metaconid and paraconid peaks. This area comes into wear later than the peaks themselves, the latter of which are used for measurements in Klein's crown height studies. The delay in attrition I describe here therefore could be an artifact of measuring a slow-wearing valley rather than a fast-wearing peak. However, findings in the earlier Neolithic cattle study were based on measurements of metaconid heights, yet produced similar results. Thus, while the lag in wear may indeed be longer in the saddle area, I have no substantive reason to doubt that my site of measurement has significantly affected the pattern of slow initial wear reported here.

The Bias in Age Estimates Is Both Directional and Age-Dependent

I now return to two issues raised earlier in the review of Klein *et al.*'s known-age wapiti analysis: directionality of the observed errors in age estimates and acceptable margins of error. The percent error figures show that soon *after* three years of wear on bison molars, the quadratic formula begins to produce ages that err substantially toward the old end of the age scale. Quadratic age estimates from more worn dentitions show greater variability around actual ages. If we compare these results to Klein *et al.*'s (1983) experience with their known-age wapiti sample, some sharp contrasts appear. The bison data show only a short span of time in which age estimates fall evenly about the real age. For most of the lifetime, quadratically-derived age estimates are *directionally* biased in an age-dependent way.

Recall that Klein *et al.* (e.g., 1981) state that plus or minus 10% of lifespan is an acceptable range of accuracy in ages derived from the quadratic method, based on the wapiti control study. Given the potential ecological longevity value used for bison (25 years=300 months), this could be taken to mean that, as long as the age estimates fall within plus or minus 30 months of the actual age, they are acceptable. From this point of view, much of the percent error patterning shown for samples of various ages is acceptable, since it falls within that range (167% for 18 months, 100% for 30, 71% for 42 months, 56% for 54 months, and 25% for 120 months). These facts are incontrovertible, but it remains true that an acceptable range of error depends on what the investigator is trying to learn about human behavior from the mortality data. If more accurate age estimation methods are needed, as they often are in studies of domestic animal use, the rule-of-thumb 10% pattern used to define the range of error may not be acceptable at all. This also raises the question of whether crown height can be used to predict age-at-death with acceptable levels of accuracy.

IMPLICATIONS FOR CONSTRUCTING MORTALITY PROFILES

Published mortality profiles using the quadratic method rely on age data derived from one permanent tooth and one deciduous tooth to document the total lifespan. If the lifetime is finely subdivided for analytical purposes, and if early phases of molar wear differ from tooth to tooth, which teeth to select in constructing mortality profiles is a critical consideration. I previously noted problems constructing mortality profiles for Neolithic cattle with ages derived from the quadratic formula. Comparisons of mortality profiles constructed with quadratically-derived ages for different molars raised troubling interpretive issues that would have been obscured had I simply used one molar to create the profiles. I hypothesized that the statistically significant differences among the profiles resulted from biologically-based attrition patterns in the cattle, rather than from selective slaughtering of two or three separate age classes by ancient herders. This is, I stress, a problem with age profiles that are finely subdivided (e.g., 10% of lifespan or by year in long-lived species).

How does the choice of molar affect the shape of the resultant mortality profile? Figures 4.11 and 4.12 show grouped mortality profiles of the Wheatland and Yellowstone lower molar samples, grouped by 12-month intervals beginning at 6 months of age. At the top of each figure is the actual age structure of the sample, below are profiles constructed from quadratic formula age estimates for each molar. The impact of the molar-sparing phenomenon is clearly visible in the samples. For each molar, profiles constructed from quadratic ages show a strong bias toward determinations close to age-at-eruption (6 to 30 months), despite the fact that actual ages-at-death are more evenly distributed from 6 to 54 months in both samples. For older animals, quadratic age determinations are more widely distributed than are actual ages. Readers are reminded that the spans of time over which attrition is monitored by the actual bison cohorts varies. For M1's, up to nine years' wear (age minus age at eruption) is represented in the Yellowstone sample, while the Wheatland sample derives from only the first four years' attrition. For M2's, the attrition spans monitored are up to eight years (Yellowstone) and to three (Wheatland). For M3's, actual attrition spans represented are seven years (Yellowstone) and two years (Wheatland).

One might initially imagine sidestepping the problem of anterior molars being "too young" by selecting the last-erupting M3, but the M3 age estimate distributions indicate that this strategy may not solve the problem. Neither the Wheatland nor the Yellowstone samples are actually dominated by animals that are really around 30 months old. Animals in the 30-month age class predominate in the Wheatland M3 mortality profile derived from the quadratic estimates, however, and the Yellowstone estimates produce a similar pattern. In each case, the peak in 30-month age determinations is created by older, adjacent age classes contributing spuriously young, near age-at-eruption estimates.

Nearly all the derived profiles differ significantly from the actual age profile for their respective samples and from one another. Table 4.4 summarizes Kolmogorov-Smirnov statistics for each sample. Quadratically derived mortality profiles for the Wheatland and Yellowstone samples differ from the actual age profiles at the $p<0.05$ level, with one exception (Yellowstone M1:Actual Age). Differences are even more pronounced for profiles derived from different molars of the same mandibles. Only one comparison of quadratic-derived profiles, Wheatland M2:M3, produced a statistic indicating that the molar samples could have been drawn from the same population. In sum, mortality profiles constructed from quadratic formula ages on different molars of a single sample

70

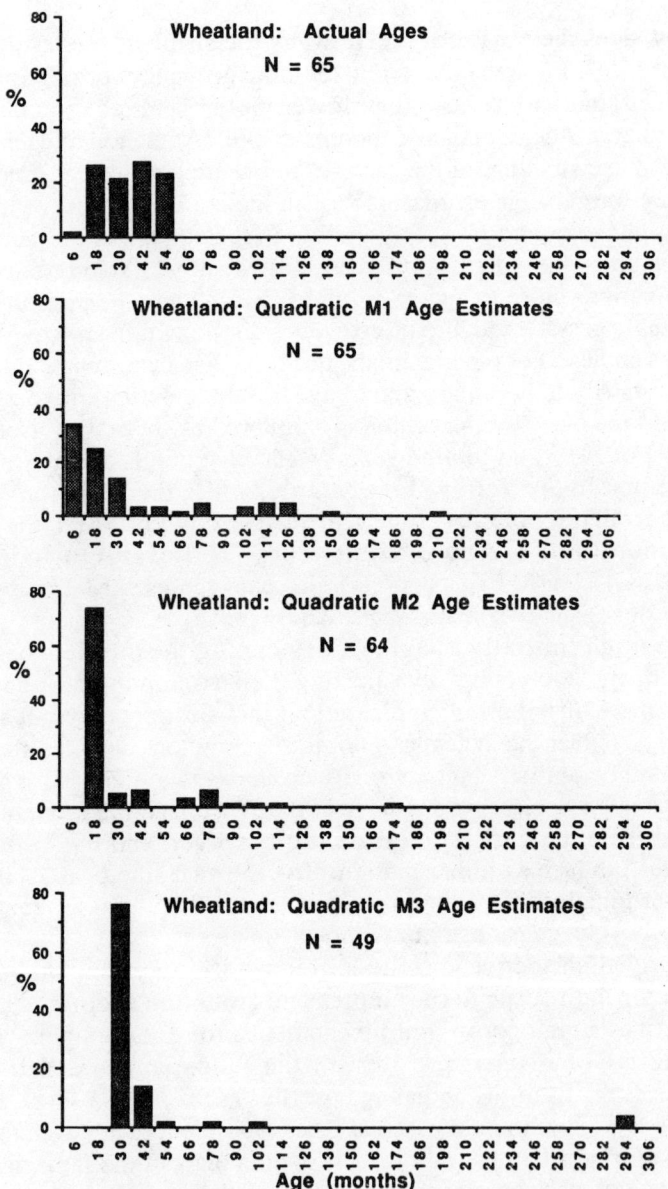

Figure 4.11. Mortality profiles for Wheatland sample: top profile is constructed from actual ages, others from quadratic-derived estimates from the three molars.

71

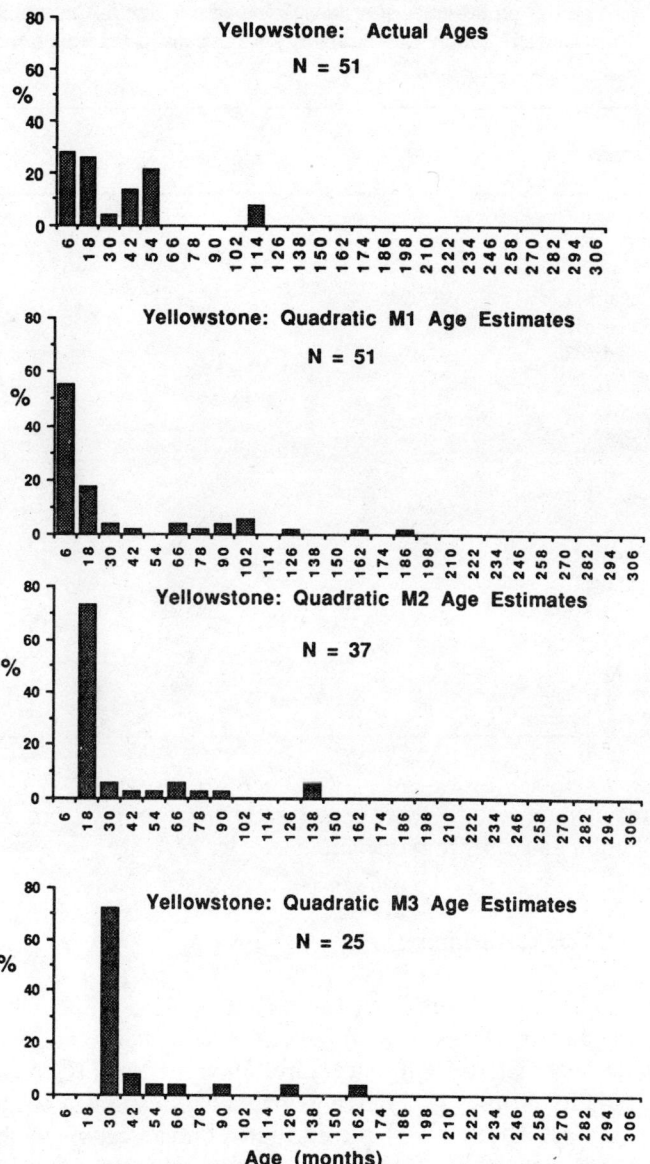

Figure 4.12. Mortality profiles for Yellowstone sample: top profile is constructed from actual ages, others from quadratic-derived estimates from the three molars.

Table 4.4. Summary of probabilities of results, Kolomogorov-Smirnov tests comparing Wheatland and Yellowstone mortality profiles constructed from actual ages at death (Figs. 4.11 and 4.12).

Curves Compared	p
Wheatland:	
Actual:M1	.03
Actual:M2	.004
Actual:M3	.017
M1:M2	.027
M1:M3	.5
M2:M3	.001
Yellowstone:	
Actual:M1	.083
Actual:M2	.052
Actual:M3	.015
M1:M2	.002
M1:M3	.003
M2:M3	.006

produce results that tend to differ consistently and strongly from the actual age distribution and also from one another.

Catastrophic Versus Attritional?

Klein and his co-workers (e.g., Klein 1978, 1982a, 1982b; Klein et al. 1983; Klein and Cruz-Uribe 1983a) have stressed the differences between catastrophic (also called living-structure) and attritional (also called U-shaped) mortality profiles for inferring ungulate procurement behaviors of ancient hominids. Klein et al. have attributed human causes to shapes of mortality profiles constructed from ungulate dentitions, specifically certain kinds of hunting as opposed to scavenging (e.g., Klein 1978; Klein and Cruz-Uribe 1983a). These inferences have been challenged on several grounds, including lack of independent, supporting evidence for hominid roles in creating the dental samples studied and lack of attention to effects of selective transport (Binford 1984), as well as the possibility that similar profiles can arise from different agencies and processes (Stiner 1990b; also Chapter 8 of this volume).

Here I raise another related issue: the possible influence of the molar-sparing phenomenon on mortality profiles constructed from the quadratic method. Both types of curves noted above feature high frequencies of deaths in younger age classes (see also Chapter 1). Some of the quadratic-derived mortality profiles in Figs. 4.11 and 4.12 lack representation of the youngest age classes that would be provided by adding ages from a deciduous premolar. However, the impact of their strong bias toward the age-at-eruption age class gives them overall shapes which might, depending on the input from deciduous premolars, approximate either catastrophic or attritional profiles. The actual age profiles for the Wheatland and Yellowstone samples look nothing like these.

To summarize, I believe the bison study implies that all mortality profiles constructed for bovines from the quadratic formula should be reevaluated, because the overall mortality profiles might be distorted by the method. Even if this does not prove to be the case, reevaluation is bound to reveal finer variation among assemblages that is of scientific interest. Especially crucial are the "catastrophic" profiles with many "young" teeth. The cattle and bison data suggest that such profiles may reflect the dentitions of animals well into their reproductive years and of near-maximum body size. Similar caution may be appropriate for all other quadratically-derived profiles from teeth of hyper-hypsodont taxa.

CAN CROWN HEIGHTS BE USED TO AGE TEETH RELIABLY?

In view of the foregoing results, it is reasonable to ask whether crown heights can be used at all to age teeth reliably at a fine-grained scale, with either 10%-of-lifespan or calendrical intervals. Preliminary curve-fitting with the known-age samples indicates that it is possible to devise a model of molar attrition for the known-age bison samples and to derive reliable age estimates from crown heights (Gifford-Gonzalez n.d.a). These preliminary observations support Klein's and Spinage's idea that rates of dental attrition vary over the lifetime of an animal, but they suggest a *different pattern* of attrition.

CAN INITIAL CROWN HEIGHT VARIABILITY BE STANDARDIZED BY DIVIDING BY BREADTH?

The known-age bison study shows that initial crown heights of any given molar may vary as much as 10 mm (Tables 4.3 a, b and c). This is only one of many examples of the variation noted by numerous researchers, including Ducos (1968), Lowe (1967), and Klein (Klein *et al*. 1983). These authors considered that such size-related variations in crown height could skew age estimates derived from teeth. To cope with this possibility, Ducos (1968) introduced a correction factor into his study of Neolithic cattle mortality. Before deriving correlations of crown heights to known-age in a sample of West African N'Dama cattle, he divided each crown height by basal crown breadth, creating an "index of wear" (Ducos 1968:10). Although it makes good intuitive sense that crown height is related to size, the possibility of a strong relationship between initial crown height and molar breadth or any other related measure of size has not been fully explored.

To assess this, I examined the relationship between initial, unworn crown height (CHo) and basal enamel breadth in the bison sample. Figure 4.13 (top) indicates not significant relationship between these variables for M2. (First lower molars were represented by only four unworn specimens and therefore were not included.) The data points for M3's (bottom) are less dispersed along the vertical (crown height) axis, but can be seen to differ substantially with respect to the horizontal (basal breadth) axis from the pattern shown by M2's in the same sample. I believe these differences in both dispersion of points and the r values counsel caution in generalizing about the strength of this relationship in cattle. Researchers would do well to investigate this relationship in any sample with which they work before deciding to apply a correction factor.

This point is reinforced by two other studies I conducted on unworn molars: one of Neolithic East African cattle and one of 100 caribou mandibles from the historic Nunamiut site of Palangana (see also Binford 1978, 1981). Among the cattle molars assessed, no consistent relationship emerged between crown height at eruption and basal breadth. Pearson's r values ranged from -0.24 to +0.12 for different unworn molars (sample sizes ranging from 10 to 13). Crown heights at eruption for caribou lower M2's and M3's are known only for the Palangana sample, and these manifested no strong relationship between crown heights and basal breadths (sample sizes of 10 and 13, respectively).

These results on artiodactyl species in different families represent considerable differences in degree of hypsodonty and suggest that there is

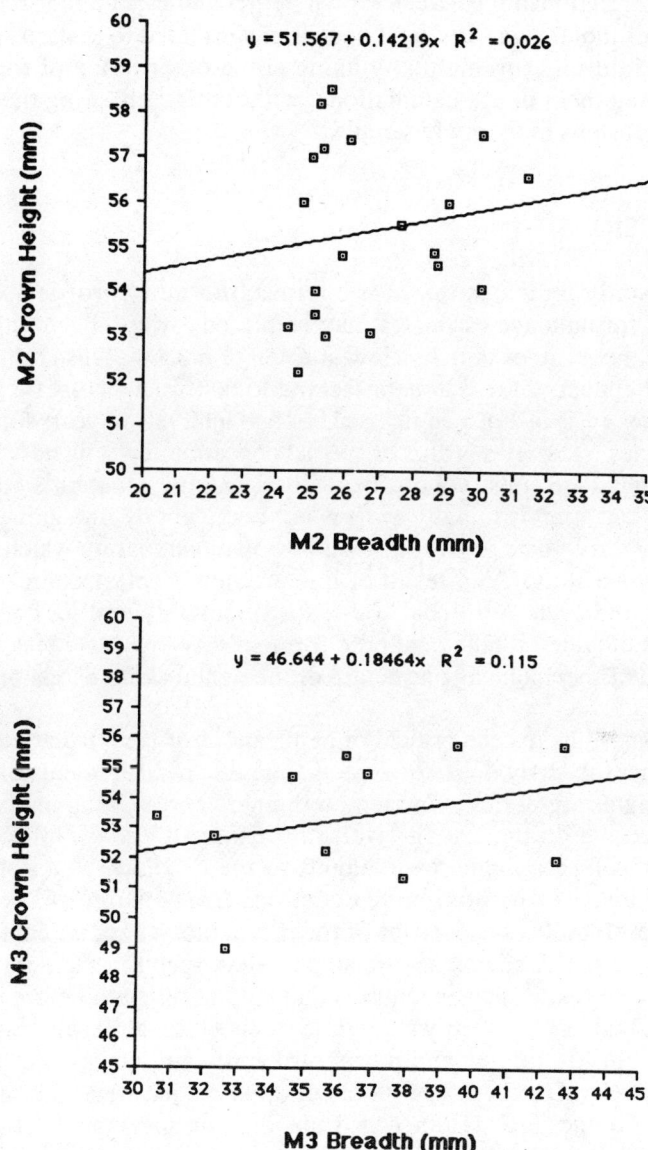

Figure 4.13. Relation of basal enamel breadth to initial (unworn) crown height, combined Wheatland (N=1) and Yellowstone samples (N=15).

no regular relationship between crown height and basal molar breadth as an index of molar size. It would appear unwarranted to seek to "correct" crown height measurements by using some other index of tooth size before using them in age calculations, without first checking these basic size relationships in the study sample.

CONCLUSIONS

This study indicates that, for bovines, mortality profiles based on quadratic formula age estimates may be biased toward the youngest age classes as the result of initially slow attrition of molars. This problem is of particular concern to researchers seeking to construct fine-scale profiles, using either 10%-of-lifespan intervals or age intervals in years for longer-lived species. This bias is also of special concern to investigators who use isolated molars to construct profiles. The quadratic equation's prediction that dental attrition is initially swift is not borne out by fine-grained study of the first three years of wear in each bovine molar, during which attrition appears to be slow. As a result of this problem, "catastrophic" (living-structure) and even "attritional" (U-shaped) mortality profiles constructed with ages obtained by the quadratic formula may provide a less accurate reflection of the actual age structure of the dental sample than originally assumed.

Although I have been critical of applications of the quadratic formula, I stress that the study does not demonstrate that measurements of enamel crown heights are useless. In fact, preliminary curve-fitting indicates that crown heights do predict age well in the known-age samples. Better results in using remnant crown heights to age ungulates can probably be obtained by deriving predictive equations from studies of known-age species, preferably conspecifics of the archaeological or paleontological taxon under study, or a closely related modern species. I am not arguing that every ungulate species requires this kind of research before it can be aged. Rather, I believe it worthwhile to check on a representative of a group of closely related and morphologically similar taxa (e.g., bovid tribes), given differences in patterns of eruption and wear. For example, although bovines are high-crowned bovids, it is advisable to study wear patterns among representatives of other hyper-hypsodont bovid tribes (e.g., Caprini, Alcelaphini) which begin to wear the tops of their molars before the bases have fully formed. This research would provide more specific models of dental attrition, in turn yielding more precise estimates of age.

This study raises another methodological point. Mortality profile analysis requires a thorough exploration of the possible impacts of all factors that may cause variability among age determinations from different teeth of a single species sample. Systematic checks of age profiles from all molars can reveal both recovery biases and intrinsic biologically-based biases, even in the absence of a highly precise aging method. I suggest that this be a regular step in mortality data analysis, regardless of the aging method used.

Klein and his coworkers are correct in stating that enamel crown heights are workable means for estimating age-at-death. However, some refinements of the methods originally proposed and the means for checking them are needed. This is especially true when detailed information on relatively fine subdivisions of the lifespan are sought. If these steps are taken, I believe that archaeologists and paleontologists can have considerable confidence in this approach to age estimates from ungulate teeth and in the mortality profiles based upon it.

ACKNOWLEDGMENTS

Research with Neolithic cattle dentitions was supported by National Science Foundation grant BNS-821152, under authority to conduct research from the Office of the President of the Republic of Kenya. Access to the collections was granted by the Trustees of the National Museums of Kenya. Preliminary analysis of these data was undertaken while a Visiting Scholar at the School of American Research, Santa Fe, New Mexico.

Research on the bison dentitions was sponsored by a grant from the National Science Foundation Research Opportunities for Women Program and a research grant from the Division of Social Science, University of California, Santa Cruz. The bison specimens were made available through the help of Danny Walker (Wyoming State Archaeologist), and with the permission of Mark Boyce (Department of Zoology and Physiology). Many thanks to George Frison (Department of Anthropology, University of Wyoming) for his help in locating the collections. I also thank Lewis Binford (Department of Anthropology, University of New Mexico) for allowing me to study the Palangana mandibles.

I thank those who have helped me work with the crown height data: David Barry, Henry Harpending, and Jennifer Morris. Mary Meagher (Research Biologist, Yellowstone National Park) supplied me with valuable information about the Yellowstone herd. I thank those who have read and commented on earlier drafts of this work: Lewis Binford, Jane Buikstra, Carolyn Clark, Don Grayson, John Speth, Mary Stiner, and Anna Tsing.

I owe special debts to Karen Lupo, whose scrupulous data collection provided a sound base with which to work, to Fiona Marshall, for her personal support during the cattle project, and to Richard Klein, who has generously shared unpublished data and commented on earlier drafts.

NOTE

1. Measurements of these dental samples were collected by Karen Lupo, Department of Anthropology, University of Utah.

5

Procurement Technology and Prey Mortality Among Indigenous Neotropical Hunters

Michael Alvard and Hillard Kaplan

INTRODUCTION

The elaborate and systematic use of tools in the food quest is often considered to be a distinctive feature of the genus *Homo*. In fact, archaeologists define stages in human evolution by the tool technologies found at sites of varying ages and expect changing technologies to be associated with changing subsistence regimes. However, there have been few comparative ethnographic studies of the relationship between tool technologies and subsistence patterns that can guide inferences about past hominid behavior (for exceptions, see Binford 1978, Hill and Hawkes 1983, Yellen 1977, Yost and Kelley 1983, Kuchikura 1988).

In order to understand the implications of changes in technologies found in the archaeological record for hominid subsistence regimes, we need to know how alternative technologies affect efficiency in predation strategies and how those changes might lead to differential mortality patterns in prey death assemblages. There is also the closely related issue of conservation in modern situations, particularly the impact of nontraditional technology on indigenous ecological systems (see Kaplan and Kopischke 1991). For example, the available data suggest that hunters using shotguns often ignore the small game normally pursued by bow-and-arrow hunters in the same environments (Hill and Hawkes 1983). This pattern appears to result from the fact that shotguns increase the efficiency with which large game are captured, thereby rendering the pursuit of small game less productive than the search for larger animals. This difference in hunting strategies is reflected in the quantities of small and large prey killed by bow and shotgun hunters.

Among nonhuman predators, there is some evidence that differences in pursuit strategies not only affect the species composition of the diet but also the age-structure of prey taken. For example, Bertram (1979) suggests that predation strategies fall into two general classes, defined by the length of pursuit after encounter. Ambush hunters, which engage in short pursuits, are not selective with respect to the age and sex of their prey and tend to produce mortality patterns that resemble the age-structure of the living populations. Cursorial predators, such as dogs and hyenas, chase prey over longer distances and produce a mortality pattern that over-represents younger and older age grades relative to their abundance in living populations.

Stiner (1990b) has attempted to apply the same basic classification scheme to death assemblages produced by ancient hominids as a method for learning about prehistoric subsistence behaviors and some general features of hominid predatory niche (see also, Chapter 8 of this volume). Stiner identifies a shift within the Upper Pleistocene from a highly variable set of procurement patterns to one that tends to focus on prime adult prey. She points out that this could happen *either* as a result of increased use of projectile weapons which permit a greater distance between hunter and prey *or* simply greater cooperation among hunters, as suggested by the late Mousterian archaeological record. Investigations of ancient hominid foragers can profit from knowledge about variation in subsistence practices of modern humans because some of the same general principles may be operative throughout a broad range of environmental and technological conditions. With this in mind, we present observations about the relationship between tools, pursuit tactics and the resulting mortality patterns produced by living human populations using varying technologies in the tropical forests of Peru. While the use of shotguns obviously has no direct application for prehistoric lifeways, we suggest that strategic differences apparent between shotgun and bow hunting contexts are useful for modeling differences in kill power between free-flying projectiles and hand-held weapons.

In this chapter, we specifically compare the predatory strategies of bow and arrow hunters with those of neighboring shotgun hunters in two native hunter-horticulturist communities in the lowland tropics of southeastern Peru. We examine how hunting strategies vary between the two technologies with respect to pursuit time and prey selection by species, age and sex. Evidence will be offered to show that (1) hunters using both technologies take terrestrial prey in proportion to the likely age and sex structure of the living populations, (2) both sets of hunters target adult primates, (3) the sex ratio of primate kills is female-biased among bow hunters and male-biased among shotgun hunters, and (4) hunters may respond to fat content in their choice of prey. These results are examined from the perspective of optimal foraging theory and implications for prehistoric humans are briefly examined.

THE STUDY POPULATIONS

The data presented below were collected in the Machiguenga community of Yomiwato and the Piro community of Diamante between September 1988 and June 1989. Yomiwato is inhabited by approximately 100 Machiguenga and is located on a small tributary of the Manu river in an area of upland tropical rainforest abutting the foothills of the Andes (Fig. 5.1). The Machiguenga of Manu were brought into peaceful contact with the outside world by the Summer Institute of Linguistics in the mid-1960s. In 1973, almost the entire Manu drainage was set aside as a National Park to preserve local flora and fauna, and the Machiguenga were among the traditional native inhabitants of the region allowed to remain in the Park after its establishment. Shotgun hunting was prohibited, as were extractive commercial activities. The Machiguenga currently practice an entirely subsistence-based economy, and cultivation of manioc and plantains supplies the bulk (80%) of the calories consumed. Bow and arrow hunting and fishing with a variety of technologies (bow and arrow, plant poisons and hook and line) provide 14% of the calories, and most of the protein and fat, in their diet. The remaining 6% of calories consumed are acquired by gathering fruits and insect larvae.

Approximately 200 Piro live in the village of Diamante, 90 kilometers southeast of Yomiwato. Diamante is located just outside Manu National Park on the Alto Madre de Dios, a fast, braided river of which the Manu is a tributary. The Piro of Diamante have had a longer history of contact with the modern Peruvian culture, and since they live outside the Park's boundaries, they are not subject to its regulations. As a result, Diamante Piro have adopted a number of non-traditional technologies. Some men own small 16 horsepower motors for their dugout canoes and seven men own shotguns which they commonly use for hunting. While bows are still used by the Piro, 87% of the meat acquired from hunting during our field stay was procured with shotguns. The shotguns used by the Piro are 16-gauge, single shot breechloaders obtained over the course of the last 15

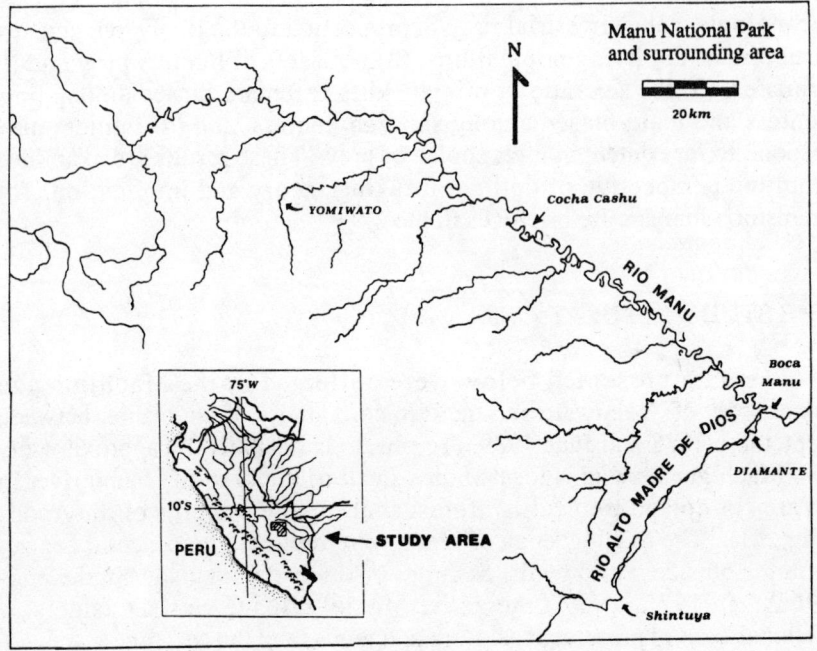

Figure 5.1. Map of the Manu National Park area.

years. The shells, containing lead shot, are obtained from river traders. Only about 5% of the calories eaten in Diamante were acquired through the purchase of commercial foods, such as rice, flour, sugar and alcohol. The remaining 95% of their calories were obtained through swidden cultivation, hunting, fishing, and collecting in roughly the same proportions as among the Machiguenga.

LOCAL ECOLOGY

About half way between the two communities, on the Manu river, is the biological research station of Cocha Cashu. Ecological studies on the flora and fauna of the area have been continuous since 1973 and the area is among the best known tropical forest regions in the New World (e.g., Terborgh 1983). The Manu and the Alto Madre de Dios rivers flow through tropical moist forest (Holdridge 1967). Along the river courses, the forest consists of successional stages produced by the river's meanderings. A number of small tributaries flow into both rivers; the community of Yomiwato is situated on one such stream. There is a marked wet (October-April) and dry season (May-September), although the onset and duration of the seasons varies somewhat between years. The rains produce a significant rise in water volume in the water courses and these seasonal fluctuations pattern many of the subsistence activities of the Machiguenga and Piro. The high and murky waters of the wet season make fishing more difficult, and a pronounced shift to highground hunting occurs during the wet season (data forthcoming). Both the Machiguenga and Piro report that primate species gain fat as the rainy season progresses making them more desirable prey, whereas they are not worth pursuing during the dry season.

Species diversity is very high in the Manu region. Terborgh *et al.* (1984) report that there are 99 species of mammals, including 13 species of nonhuman primates in the forest around Cocha Cashu. While many of these species are hunted by the Piro and the Machiguenga, as will be discussed below, only a few contributed significantly to the total number of animals killed or the total amount of meat harvested during the study period.

METHODS

The data presented below were collected using two methods. A modified form of "focal individual" sampling (Altman 1974) was employed to directly observe hunting behaviors. When a focal individual hunting day was scheduled, one of the authors (Alvard in Diamante, Kaplan in Yomiwato) would survey the adult men in the community to learn who would be hunting and select a man to follow. During focal follows, all activities of the hunter, including search, encounters with and pursuits of prey were recorded and timed. A detailed record of the location of each encounter, the prey species involved, hunting group size and the outcome of the encounter was also maintained. Encounters were defined by the hunter's report of his awareness of a potential prey animal or by his behavior. Pursuits were defined as any change in behavior, ranging from searching for to active attempts to track, chase or shoot the animal. The lengths of pursuits were measured from encounter to kill, or until an unsuccessful pursuit was terminated. The following information was obtained for each individual animal killed: species, sex, age, reproductive status, weight, and the hunting technology used. Animals encountered, or encountered and pursued but not killed, were also identified to the extent possible.

In order to obtain a larger sample of kills, information on prey taken during unobserved hunts was obtained during our visits to households. All households were surveyed, either every day or every third day, depending upon the schedule for other research activities, to determine all large game (>5 kg) killed since the last visit. In this context, the following information was elicited for each prey item: hunting method, prey species, sex, age and the location of the kill. Information on age and sex were not always consistent or reliable, and questionable cases are excluded from the sample presented here.

HUNTING TECHNIQUES AND PREY SELECTION

The majority of Piro and Machiguenga hunts involve nonspecific search for game and opportunistic encounters. Animals are encountered by sight and/or sound, sometimes flushed by dogs if present, and for some species, lured into ambushes by imitating the animal's vocalizations. Although men may have some idea of the prey species they are likely to encounter and pursue on a given day, they are frequently prepared to encounter any of several species. In fact, on most excursions, they carried several different items of technology such as bow and arrows or a shotgun, fish poison, and hook-and-line, in order to take advantage of all opportunities to obtain meat. Natural mineral licks were also visited often because they attract many species that can be killed as they eat the dirt. Hunts on trails of known length from Diamante and from Yomiwato were never longer than 20 km round-trip.

There are, however, some important differences between Piro and Machiguenga hunting strategies. Piro hunted in smaller parties than the Machiguenga. Observed Piro hunts included a mean of 1.4 hunters, while 2.1 men participated, on average, during Machiguenga hunts (student's t-test, $p=0.0053$; number of hunts observed are 30 and 16 respectively). Of the Piro hunts, 38% involved two men, and in 55% of those cases, one man carried a bow while the other carried a shotgun. In 75% of the observed Machiguenga hunts, two or more hunters participated. Dogs, while kept by both groups, did not commonly accompany shotgun hunters; only 3% of observed shotgun hunts also involved dogs. Dogs are a common feature of Machiguenga bow hunting and participated in 75% of hunts observed.

Other important similarities and differences between Piro and Machiguenga hunting strategies can be seen in species by species comparisons of capture techniques and kill patterns. Tapir (*Tapirus terrestris*), the largest mammal native to South America, provided the most meat in both communities, even though relatively few individuals were killed (6 and 13, in Yomiwato and Diamante, respectively). Little is known about the biology and behavior of tapir, except that they are nocturnal and solitary outside of the mother/offspring association and tend to walk the same trails consistently (Eisenberg 1989). Adults are reported to weigh between 200 to 300 kg (Eisenberg 1989), but none of the kills we observed weighed more than 155 kg and our informants claimed that they were "large" mature adults.

Although they are of similar economic importance in Yomiwato and Diamante, tapirs are taken by different techniques by the two communities. Piro hunters of Diamante take advantage of the habitual nocturnal travel patterns of tapir to ambush and kill them. Although a few encounters with tapir occurred during daylight searches, the majority were obtained by waiting at night with shotguns and flashlights or with trip-line traps placed along the tapir trails. One Diamante man actually specialized in night ambushes of tapir, accounting for 7 of the 13 kills. The Machiguenga, in contrast, have no access to flashlights and shotguns, and were not observed to use either ambush or trap techniques to kill tapirs. Instead, Yomiwato Machiguenga took tapirs using only bow-and-arrow in the context of opportunistic encounters during generalized day-time searches, active tracking and/or with tapir calls. Men who engaged in targeted searches for tapir left their dogs behind; Machiguenga hunters report that dogs are a often a hindrance because they flush prey before a shot can be fired. They also claim that dogs are unable to pursue and trap tapirs, although if one is wounded but escapes, the hunter may return home and track the wounded animal with dogs the next day. One tapir killed during the field session was obtained in this manner.

Collared peccaries (*Tayassu tajacu*) provided the second highest proportion of the meat in both communities. Collared peccaries were killed primarily with shotguns in Diamante through opportunistic encounters during generalized searches or at mineral licks. Encounters were often with groups of prey either seen or heard foraging. After the first shot was fired, remaining members of the peccary group fled. Because the animals often moved only a short distance, a second shot was frequently possible. Had dogs been present in such cases, the likelihood of a successful second shot would have been remote (see also general discussion by Frison, Chapter 2).

In Yomiwato, 47% of the peccaries were killed with bow-and-arrow using the same search pattern noted in Diamante. The remaining 53% were killed with the assistance of dogs that tracked the peccaries during long, rapid pursuits. For peccaries and several other terrestrial species, such as pacas, armadillos, capybaras and agoutis, dogs and humans proved a formidable predatory team. In some instances, one or two men generally leave camp with one or more dogs. The dogs run circles around the hunters as they walk the trails, frequently flushing animals from their foraging sites. Alternatively, the men will find tracks that appear fresh, and send the dogs off sniffing. When game are flushed, the dogs chase the animals as the hunters lag behind. The men maintain contact with the dogs by calling and in turn are answered by barks. The dogs either catch and kill the game or chase it into a hole or hollow log from which the hunters can extract it.

White-lipped peccaries (*Tayassu pecari*) are known in the Manu region but were not killed in great numbers. Although closely related, the two species differ in both body size and social organization. While there is much variation in published reports on adult body weight, the white-lipped peccary (weighing 25-35 kgs) tends to be about 10 kg heavier than the collared peccary (Kiltie and Terborgh 1983). The white-lipped peccary also travels in larger groups (50 to 200 individuals) over a much larger range than the collared peccary (usually less than 15 individuals). The Piro claim that white-lipped peccaries were present in the area in the past but none had been killed in this vicinity in the last few years. Only two white-lipped peccaries were killed in Yomiwato during the field session.

Two large rodents, capybara and paca, were taken frequently by hunters. Capybaras (*Hydrochaeris hydrochaeris*) can weigh up to 50 kg, favor aquatic habitats, and were third by weight in importance for Piro hunters. In contrast, only one capybara was killed by the Machiguenga during the field period. This difference may due to the fact that the village of Yomiwato is located on a small stream and therefore not directly associated with riverine habitat favored by the capybara. This prey species were never observed during hunts into the forest, whereas they were seen during travel in canoes along the water courses. Like tapir, capybara were frequently taken by Diamante Piro during nocturnal ambush hunts with shotguns. The paca (*Agouti paca*) is smaller rodent species weighing 5-13 kg that also tends to live along streams (Emmons 1990) and is common in the Yomiwato area. Every paca killed by Yomiwato Machiguenga was taken with the assistance of dogs.

The habits of brocket deer (*Mazama americana*), an important prey item in Diamante, are not well known (Branan and Marchinton 1987), but it is clear that they are solitary. Deer rank fourth in importance in the Piro diet by weight and were taken during opportunistic encounters with the use of shotguns. No deer were killed by Yomiwato Machiguenga bow hunters, although several were pursued.

At least seven of the thirteen species of primates found in the Manu region are hunted by the Machiguenga and the Piro (see Alvard and Kaplan n.d.). For the Machiguenga, spider (*Ateles paniscus*, weighing 6-10 kg) and woolly monkeys (*Lagothrix lagothricha*, weighing 8-10 kg) rank third and fourth, respectively, in the proportion of meat they provide in the diet. These two species were much less important for the Piro, as spider monkeys were taken only occasionally and no woolly monkeys were killed in Diamante. According to Piro reports and observed behavior, the paucity of these monkeys in the diet was due directly to availability rather to decisions not to pursue them. Howler monkeys (*Alouatta seniculus*, weighing 5-8 kg) are taken in both vicinities, although they do no contribute significantly to the total amount of meat taken in either locale. In the Yomiwato area, Machiguenga hunters rarely pursued capuchin monkeys, (*Cebus appella* and *Cebus albifrons*, weighing 2-5 kg) and the squirrel monkeys (*Saimiri sciureus*, weighing 0.6-0.9 kg). Capuchins were encountered frequently by the Piro, but were inconsistently pursued. Piro hunters reported that they generally ignored them during the dry and the early wet seasons, when the monkeys are not desirable because of low body-fat content. Squirrel monkeys were encountered often near Diamante. They were never pursued by shotgun-bearing Piro, but *were* pursued and killed by bow-hunting Piro.

RETURN RATES AND HANDLING TIMES

A number of other researchers have noted the marked advantage shotguns bring to the meat quest among neotropical hunters (Hames 1979, Hames and Vickers 1982, Yost and Kelly 1983, Hill and Hawkes 1983) and a similar pattern is apparent with the Piro. Table 5.1 compares the results of observed Piro shotgun hunts with Machiguenga bow hunts. Including all travel, search, and pursuit time, these data show that the Piro obtained 1.2 kg (undressed) of meat per hour hunting with shotguns,[1] while the Machiguenga obtained only 0.12 kg of meat per hour. This tenfold difference may be a slight overestimate due to sample error, given the small number of observed hunts, but the advantage afforded by shotguns is clear.

Shotguns are more effective because of the greater likelihood of hitting the target and greater killing power per shot. Shotgun pursuits often consist of nothing more than lining-up a shot and squeezing the trigger.

Table 5.1. Comparison of hunting efficiency between shotguns and bows.

	village Diamante (shotgun)	Yomiwato (bow)
Hours hunting observed	212	85
Mean number hunters per hunt[a]	1.4	2.12
Bow/shotgun hours observed	243	208
Number prey animals killed	39 [b]	3 [c]
Number shots	50 [b]	89 [b]
Number of shots per kill	1.3	30.0
Kg live weight harvested	291.0	24.4
Kg harvested per bow/shotgun hour	1.2	0.12

[a] Includes individuals armed with bows that accompanied shotgun hunters.
[b] Includes bird pursued and killed, but not included in Table 5.3.
[c] Includes one night monkey (*Aotus* spp.).

Even with relatively poor aim, the spread of lead shot is often sufficient to mortally wound the targeted animal. In contrast, arrows frequently miss their mark, sometimes because vegetation deflects the arrow and because accuracy decreases considerably at long distances. In addition, animals are frequently wounded with arrows but not severely enough to prevent their escape.

The disadvantage of bows compared to shotguns is amply demonstrated by the large number of bow shots taken during observed hunts by the Machiguenga and the small number of prey actually obtained. Tables 5.2a and 5.2b list data on observed pursuits of large terrestrial game and primates in the vicinities of Yomiwato and Diamante. For Piro shotgun hunters, 78% of their shots resulted in killing the prey, whereas only 3% of shots taken by Machiguenga bow hunters killed an animal. Among Diamante Piro, 15 of the 27 shotgun pursuits (55%) were successful, while only 2 of the 25 (4%) bow pursuits resulted in kills in Yomiwato.

One effect of the difference in killing power between the two technologies is that bow hunters engage in longer pursuits than shotgun hunters. Among the Machiguenga, if the hunter misses a shot, he may continue to chase the animals for several hours, sometimes shooting many arrows in the process. Moreover, the use of dogs among the Machiguenga

Table 5.2a. Shotgun pursuits of large terrestrial mammals and arboreal primates during hunts observed near Diamante.

case	number of hunters[a]	prey	mode of encounter[b]	hours pursued[c]	reason pursuit terminated	number of shots	outcome
1	1 shotgun	capuchin	called	0.37	escaped	0	none
2	1 shotgun	capuchin	called	0.60	escaped	0	none
3	1 shotgun	peccary	heard	0.18	killed	1	1 peccary
4	1 shotgun & 1 bow	peccary	saw	0.02	killed	1	1 peccary
5	1 shotgun & 1 bow	capuchin	saw	0.08	showed little interest	0	none
6	1 shotgun & 1 bow	howler	heard	0.21	killed	3	3 howlers
7	2 shotguns	peccary	heard	0.10	killed	3	3 peccaries
8	2 shotguns	spider	heard	0.03	killed	1	1 spider
9	1 shotgun & 1 bow	peccary	heard	0.02	killed	2	2 peccaries
10	1 shotgun	tapir	heard	0.78	escaped	0	none
11	1 shotgun	tapir	heard	0.57	escaped	0	none
12	1 shotgun & 1 bow	capuchin	called	0.60	killed	1	1 capuchin
13	1 shotgun	capuchin	called	0.43	killed	1	1 capuchin
14	1 shotgun	peccary	heard	0.39	escaped	0	none
15	1 shotgun	capuchin	called	0.07	killed	1	1 capuchin
16	1 shotgun	capuchin	heard	0.42	escaped	0	none
17	1 shotgun	peccary	heard	0.17	escaped	0	none
18	1 shotgun	peccary	saw	0.03	killed	1	1 peccary
19	1 shotgun	peccary	saw	3.98[d]	escaped	1	1 peccary
20	1 shotgun	peccary	heard	0.38	killed	2	1 peccary
21	1 shotgun	capuchin	called	0.25	escaped	0	none

(continued)

Table 5.2a. (Continued)

case	number of hunters[a]	prey	mode of encounter[b]	hours pursued[c]	reason pursuit terminated	number of shots	outcome
22	1 shotgun	spider	heard	0.68	killed	3	1 spider
23	1 shotgun	capuchin	called	1.00	escaped	0	none
24	1 shotgun	capuchin	heard	0.45	escaped	0	none
25	1 shotgun & 1 bow	spider	called	1.02	killed	5	2 spider
26	2 shotguns	capuchin	saw	0.17	killed	4	3 capuchins
27	2 shotguns	capuchin	heard	0.28	escaped	0	none

peccary = collared peccary; capuchin, howler, spider, woolly = monkey.

[a] Number of hunters, weapons used and presence/absence of dogs (flushed).

[b] Means by which prey was detected. Capuchin monkeys (and sometimes other prey) are often detected by imitating an animal's vocalizations and listening for a response. Piro hunters could frequently lure an entire capuchin troop into an ambush by imitating an infant's distress calls. In these situations, much of the pursuit time involves waiting for the troop to come within range.

[c] Pursuits are defined as beginning at the first instant that a hunter's activity appears directed at procuring a prey animal, and not necessarily at the moment of encounter. Howler monkeys, for example, are often detected hours before hunters actually seek them out. Pursuits end when the prey animal is killed or when the hunters no longer seek to kill the animal.

[d] In this case, the pursuit occurred over the course of two days. The peccary was wounded and pursued until it entered a hole under a tree. The hunters barricaded the exits and then returned the next day with help to dig it out.

Table 5.2b. Bow pursuits of large terrestrial mammals and arboreal primates during hunts observed near Yomiwato.

case	number of hunters[a]	prey	mode of encounter[b]	hours pursued[c]	reason pursuit terminated	number of shots	outcome
1	1 bow/dogs	peccary	flushed	0.90	escaped	0	none
2	3 bows/dogs	tapir	called	0.48	escaped	0	none
3	3 bows/dog	capuchin	heard	0.08	escaped	0	none
4	2 bows/dogs	peccary	flushed	0.42	escaped	0	none
5	1 bow/dogs	peccary	saw	0.72	escaped	0	none
6	1 bow/dogs	peccary	called	0.34	could not find	0	none
7	6 bows/dogs	peccary	flushed	0.47	killed	1	1 peccary
8	2 bows/dogs	deer	flushed	1.10	escaped	0	none
9	2 bows/dogs	capuchin	saw	0.05	hit but escaped	1	none
10	3 bows/dogs	woolly	heard	0.90	escaped	11	none
11	3 bows/dogs	spider	saw	0.07	escaped	1	none
12	2 bows/dogs	peccary	saw	0.03	dogs would not chase	0	none
13	2 bows/dogs	woolly	saw	0.50	escaped	1	none
14	1 bow	peccary	saw	0.12	escaped	0	none
15	1 bow	woolly	saw	2.07	escaped	11	none
16	2 bows/dogs	spider	heard	0.58	escaped	0	none
17	2 bows/dogs	spider	heard	0.90	too windy	4	none
18	2 bows	spider	saw	2.07	hit but escaped	18	none
19	2 bows	woolly	saw	0.22	escaped	0	none
20	2 bows	spider	heard	0.53	rain	1	none
21	1 bow	capuchin	saw	0.08	escaped	0	none

(continued)

Table 5.2b. (Continued)

case	number of hunters[a]	prey	mode of encounter[b]	hours pursued[c]	reason pursuit terminated	number of shots	outcome
22	2 bows	woolly	saw	2.10	rain	15	none
23	2 bows	woolly	saw	0.55	escaped	1	none
24	2 bows	woolly	saw	1.20	escaped	2	none
25	2 bows/dogs	paca	flushed	0.26	killed	0[e]	1 paca

peccary = collared peccary; capuchin, howler, spider, woolly = monkey.

[a,b,c] as in Table 5.2a.

[e] The paca was killed by jabbing it with an arrow as the dog held it.

Table 5.3. Mean and standard deviation of pursuit times (in Diamante and Yomiwato) for terrestrial mammals and arboreal primate prey.

prey type	Diamante (shotgun)		Yomiwato (bows)		
	mean hrs.[a]	SD	mean hrs.	SD	
terrestrial mammals[b]	.203 (.264)	+.25	.509	+.35	$F = 5.3, p = .03$
primates	.316 (.416)	+.27	.748	+.75	$F = 7.3, p = .01$

[a] The number in parentheses was calculated by treating each pursuit as a single case and ignores differences in the number of animals actually taken. The p values remain less than 0.1.

[b] Does not include pursuit case 19, in which the peccary was dug out from under a tree after it was killed.

increases pursuit times because dogs frequently scare the game into flight before a shot is taken, and they are able to keep up with running prey over long distances. In contrast, most shotgun pursuits terminate after the first shot is fired.

Table 5.3 presents summary statistics for pursuit times of terrestrial and arboreal mammals near Diamante and near Yomiwato. These results include all pursuits, both successful and unsuccessful. For Diamante Piro, two numbers are presented, reflecting two different ways of handling multiple kills during a single pursuit. The first measure treats each animal taken as a separate pursuit. If multiple animals were taken during a single hunting event, the total time was then divided by the number of animals captured. For example, pursuit number 6 in Table 5.2a lasted 0.21 hours and resulted in the harvest of three howler monkeys by shotgun. This is scored as three pursuits of 0.21/3 hours or 0.07 hours each. The number in the parentheses treats each pursuit as a single case and ignores differences in the number of animals taken, and for this measure, the same howler example is scored as one pursuit lasting 0.21 hours.

A total of 13.3 hours and 16.7 hours of pursuit were observed with the shotgun and bow hunters, respectively. An analysis of variance shows that for both arboreal and terrestrial mammals shotguns significantly decrease average pursuit times ($F=7.3$, $p=0.01$ and $F=5.3$, $p=0.03$, respectively). Average bow pursuits are more than twice as long as shotgun pursuits. One kilogram of meat procured by shotguns took, on average, 0.046 hours (2.8 minutes) of pursuit to obtain, while a kilogram procured by bow took on average 0.70 hours (42 minutes), a fifteen-fold difference in weight. For both groups, pursuits of primates took somewhat longer than those for terrestrial mammals, although the difference is not statistically significant (Diamante: $p=0.26$, Yomiwato: $p=0.24$).

THE AGE STRUCTURES OF KILLS

Table 5.4 presents the age structures of total kills by species for each community and compares them to the composition of free-ranging groups of the same species, based on independent censuses. The data from Diamante and Yomiwato are derived from all kills made during the field period, not just prey taken during hunts with an investigator present. Ideally, in order to determine how Machiguenga and Piro hunters harvest prey, it is necessary to compare the age structure of kills to the actual age structure of the subpopulations being harvested. Given that the data on the age structure of those populations are unavailable, the results of studies done by biologists in other parts of the prey's range are substituted. Sowls (1984) provides a large sample from which our control data for peccary are extracted, but it should be noted that the reference populations live in the extreme northern end of this species' geographic range, in the desert environments of south Texas and Arizona. Census data on capuchin (*Cebus apella*) and spider (*Ateles paniscus*) monkeys were collected at the Cocha Cashu research station within Manu National Park (see Fig. 5.1). Reliable population structure data for many of the other prey species are not yet available.

The age structure of primate kills appears to be slightly skewed in favor of adults. Information on the living populations of primates indicate that about 60% of the individuals are adults, while fully 82% of the primates killed in Diamante and 73% of those killed in Yomiwato were adults. This bias in favor of adult prey is actually more extreme than the data suggest, because many of the subadults killed were dependent infants that died when their mothers were killed; within the subadult age category, 40% of primates taken near Diamante and 69% taken near Yomiwato were infants that were not the hunters' primary targets. If these nontargeted individuals are excluded, about 90% of the monkeys killed in both

Table 5.4. Age composition of harvested prey by species obtained by shotgun hunters near Diamante and by bow hunters near Yomiwato compared with census data for free-ranging populations.

		Diamante			Yomiwato			censused populations	
prey species	common name	N	subadult[a]	adult	N	subadult[a]	adult	subadult[a]	adult
Tapirus terrestris	tapir	7	.29	.71	6	.17	.83	--	--
Hydrochaeris h.	capybara	5	.20	.80	-	--	--	--	--
Agouti paca	paca	-	--	--	6	.0	1.0	.43	.57[b]
monkeys[c]	---	28	.18	.82	48	.27	.73	.61	.39[d]
Mazama spp.	deer	12	.17	.83	-	--	--	.59	.41[e]
Tayassu spp.	peccary	46	.26	.74	21	.24	.76	.25	.74[f]

[a] Includes all adolescents and infants.
[b] Source is Collett (1981).
[c] Monkeys include spider (*Ateles paniscus*), howler (*Alouatta seneculus*), woolly (*Lagothrix lagothricha*), and capuchin (*Cebus* spp.).
[d] Crockett and Eisenberg (1987), Janson (1984), McFarland Symington (1988), Nishimura and Izawa (1975), Izawa (1976).
[e] Branan and Marchinton (1987).
[f] Sowls (1984).

communities are adults, a 50% increase over their reported proportion in the living populations studied.

About 75% of the terrestrial mammals taken in the vicinity of the two communities were adults. This figure is close to Sowls' (1984) results for living populations of peccaries and higher than Branan and Marchinton's (1987) census figures for deer. The data are insufficient to determine whether hunters specifically targeted adult terrestrial prey. Informants report that they will pursue the large prime adult animals if they can choose among several possible targets. However, if the only possible target is a sub-adult, it will be shot, especially for the larger-sized species.

For species where samples are sufficiently large to compare prey age selection *between the two communities*, we find no significant differences in age structures of terrestrial or arboreal prey killed by hunters (peccary: Chi-square=0.039, p=0.86; tapir: Chi-square=0.258, p=0.61; primates: Chi-square=0.792, p=0.37).

KILL PATTERNS BY SEX OF PREY

Table 5.5 compares the sex ratio of kills in the two study communities with those reported for free-living populations. The sex ratios of harvested terrestrial prey species for which there are sufficient data, peccaries and deer, were not significantly different from the living populations (peccary: Chi-square=0.445, p=0.50; deer: Chi-square=0.024, p=0.88) on the part of either the Piro or the Machiguenga; that is, the sex ratios of animals killed did not differ significantly from an even (100:100) sex ratio.

Examination of the sex ratios of the primate kills reveals a different pattern. The sex ratio of the primates killed by bow are skewed strongly towards females. Among spider and woolly monkeys, females were killed more than four times as often as males, and the sex ratio of their kills was significantly different from 100:100 (spider monkey: Chi-square=5.8, p=0.02; woolly monkey: Chi-square=3.2, p=0.07). In contrast, the sex ratio of howler, spider, and capuchin monkeys killed by shotgun hunters did not differ significantly from 100:100 (howler monkey: Chi-square = 0.046, p=0.83; spider monkey: Chi-square=0.303, p=0.58; capuchin monkey: Chi-square=0, p=1).

However, census data on living groups of primates also show a female-biased sex-ratio. A number of workers have noted, for example, a skewed sex ratio for spider monkeys across the genus and in many geographic locations (Chapman *et al.* 1989), including spider monkeys in

Table 5.5. Sex of selected prey species killed near Diamante (shotguns) and near Yomiwato (bows) and sex ratios based on census data for free-ranging populations.

prey species	common name	Diamante ratio[a]	(N)	Yomiwato ratio[a]	(N)	censused populations ratio[a]
Alouatta seniculus	howler	83	(11)	100	(2)	60 [b]
Cebus spp.	capuchin	100	(16)	0	(3)	100 [c]
Tapirus terrestris	tapir	150	(10)	500	(6)	--
Agouti paca	paca	--		100	(6)	100 [d]
Hydrochaeris h.	capybara	100	(10)	--		--
Ateles paniscus	spider	150	(15)	13	(17)	35 [e]
Lagothrix lagothricha	woolly	--		33	(24)	101 [f]
Mazama spp.	brocket deer	110	(9)	--		96 [g]
Tayassu tajacu	collared peccary	87	(62)	80	(18)	112 [h]

[a] Sex ratio is based on the number of males per 100 females.
[b] Source is Crockett and Eisenberg (1987).
[c] Janson (1984).
[d] Collett (1981).
[e] McFarland Symington (1988).
[f] Nishimura and Izawa (1975), Izawa (1976).
[g] Branan and Marchinton (1987).
[h] Sowls (1984).

Manu (McFarland Symington 1988). According to the data from living populations, 63% of the spider monkeys available to hunters are in fact females. Yet, 88% of those killed by bow hunters of Yomiwato were females. This difference is not statistically significant (Chi-square=1.125, p=0.28), but the statistic is based on a small sample of kills and probably deserves further attention as larger samples become available in the future. The sex ratio of spider monkeys killed by Piro shotgun hunters is significantly skewed toward males, compared both to living populations (Chi-square=3.537, p=0.06) and those killed by bow users (Chi-square=8.2, p=0.004). This suggests that male spider monkeys are more vulnerable to shotguns and that female spider monkeys are marginally more vulnerable to bow hunters.

Woolly monkeys have not been well studied. Available data on group composition show considerable variability in sex ratios. Nishimura and Izawa (1975) report a male-biased sex ratio of 120:100 among woolly monkeys in Venezuela, whereas Izawa (1976) found a female-biased ratio of 80:100 in another group. In the absence of better alternatives, the ratio presented in Table 5.5 is the average of these two values. Yet, the sex ratio of woolly monkeys harvested by the bow hunters was 33:100. This difference from a 100:100 sex ratio is of borderline significance (*Chi*-square=3.2, p=0.07). Woolly monkeys were not encountered by shotgun hunters.

These data suggest that the terrestrial mammals are taken in proportion to their natural frequencies in living populations in both areas, while the arboreal primates display sex-specific vulnerability, depending on whether shotguns or bows were used.

DISCUSSION AND CONCLUSIONS

Technology, Dogs and Hunting Strategies

Both shotguns and bows allow hunters to target and kill animal prey from a distance. They are effective for arboreal, flying and terrestrial animals. The major difference between the two types of weapons is that shots fired from guns are more likely to result in a kill because they are more likely to strike the target and a strike from a gun is more deadly. Numbers of prey killed per encounter, killed per shot, and overall hunting return rates are all many times greater for shotgun hunters than for bow hunters.

One result of this difference in kill power is that shotgun pursuits tend to be shorter in duration than bow pursuits. Bow hunters tend to pursue their prey over greater distances and longer periods because many shots are required. This is particularly true for primates and peccaries, because they can be followed effectively by hunters, even after shots are fired. Differences in kill power and pursuit times associated with the two technologies have a number of consequences for the overall strategies employed by Piro shotgun and Machiguenga bow hunters, and the resultant patterns of mortality in their prey.

The Machiguenga hunt in larger parties than the Piro. This may be related to the difference in pursuit lengths and hunting tactics. When the Machiguenga hunt monkeys, groups of two or three men are quite common. In most monkey hunts, while one hunter searches for and gathers arrows that have missed the target, other hunters continue pursuing the quarry. Such tactics are less useful for shotgun hunters, who often kill their prey with a single shot. Similarly, Machiguenga hunting tapirs tend to hunt alone because tapirs are not easily pursued after a shot is fired.

The common use of dogs by the bow hunters and their conspicuous absence from shotgun hunts also appears to be related to differences in hunting technology. Animals such as peccaries, pacas and capybara, which tend to flee and then hide, can be followed effectively and held at bay by dogs. However, there is a trade-off in the use of dogs: they are noisy and signal the hunters' presence to the prey. Therefore, when a man hunts with a dog, it is less likely that he will be able to surprise the animal and take a shot before it flees. The Machiguenga hunting data suggest that dogs compensate for the loss of surprise with their ability to track and detain prey. For shotgun hunters who are very likely to kill an animal with the first clear shot, the use of dogs is more likely to lower their return rate. The Piro claim that before shotguns were readily available, many hunting dogs were kept.

Inter-species Prey Choice

Optimal foraging models predict optimal prey choice for animals, given explicit constraints and the nature of the suite of possible prey items (Charnov and Orians 1973). These models predict which prey items should be pursued and which ignored in order to maximize the predator's long-term net rate of energy capture. In the optimal diet model, prey are ranked according to profitability as a function of the expected net energy gain from each prey type per unit handling time. Handling refers to the time required to pursue, capture and consume a prey item. According to the model, prey items are added to the diet in order of profitability until the pursuit of the next prey type would lower the average return rate as compared to continued search for more profitable prey items.

Examination of the data indicate both the Machiguenga and Piro always pursue prey that weigh more than about 4-5 kg, which also tend to be the most profitable. Nevertheless, Machiguenga killed more tapir, and spider and woolly monkeys per consumer than the Piro. The Piro, in turn, killed more peccary, deer, capybara and capuchin monkeys than the Machiguenga. We suspect that many of these differences are a function of sampling error and local variation in prey availability between the two communities. In the case of the largest animals, such as peccary and tapir, sampling error even over the course of an entire year is likely to be important because a few kills can produce large differences in the amount obtained per consumer. In the case of deer, capybara, spider and woolly monkeys, ecological differences in prey density between the hunting zones of the two communities probably account for most of the differences. A definitive test of this hypothesis requires better information on game densities, which are now being collected.

The harvesting of capuchin monkeys provides a counter-intuitive result that may be related to the effects of shotgun and bow technologies on long-term energy maximization strategies. The smaller-bodied capuchin monkeys are common in the hunting areas of both communities. During the rainy season, when they are reportedly fatter, capuchins are actively pursued by the Piro. Machiguenga bow hunters, on the other hand, largely ignore capuchins throughout the year. It is possible that two factors interact to create this pattern. Because bow hunters engage in long, frequently unsuccessful, pursuits of primates, their returns upon encounter are low. If Machiguenga bow hunters were to pursue the smaller capuchins (which weigh only about 3 kg), their returns would be much lower upon encounter than those obtained from pursuits of spider or woolly monkeys that weigh more twice as much. Since all three species of monkey travel high in the forest canopy, the time it takes to kill them would be similar, but the returns from a successful capuchin kill would be less than half as much as for either of the other two species. It appears that Machiguenga ignore the capuchins in order to continue searching for more profitable larger-bodied animals. As long as encounters with those larger animals are sufficiently frequent, Machiguenga bow hunters may achieve higher long-term returns by ignoring the capuchins. In contrast, since Piro shotgun hunters do not engage in long multi-shot pursuits, encounters with capuchins do not result in much lost search time for bigger animals. For them, it appears that short capuchin pursuits leads to an increase in return rates, especially given the low densities of spider and woolly monkeys in their hunting area.

The opposite pattern is found among the Ache in Paraguay. Shotgun hunters ignore capuchins and bow hunters pursue them avidly. Hill and Hawkes (1983) report that the return rate for Ache shotgun hunters pursuing larger animals is just high enough that pursuits of capuchins could lower shotgun hunters' return rates. There are no spider and woolly monkeys in eastern Paraguay; the only large-bodied primates are howler monkeys and they are infrequently encountered.

Such inconsistent results between study areas suggest that the effects of technology may interact with features of local ecology. It may be difficult to derive simple generalizations about the species composition of kills, as a function of weapons used, without also taking into account the specific composition of animal communities at the level of local environments, particularly in tropical forest and desert settings.

Intra-specific Prey Choice

Simple optimal foraging models assume discrete encounters with single prey items. However, many of the prey species hunted by the Machiguenga and Piro travel in groups and are encountered simultaneously. More complex optimal foraging models treat these as encounters with a 'patch' and are designed to predict how much time will be spent exploiting the patch (Stevens and Krebs 1986). Optimal patch exploitation times are predicted on the basis of the expected return rate from exploiting the patch, which diminishes with handling time, relative to the expected returns from searching for and exploiting other patches.

Differential vulnerability of individuals within a species complicates these models. When a hunter sees or hears a troop of monkeys or a herd of peccaries, it is not simply an encounter with multiple individuals that are identical. Some individuals are larger and some are more vulnerable than others. All else being equal (such as ease of capture), the hunter should choose the individual that would be expected to return the most energy; this would be the largest individual of the group encountered. Both Machiguenga and Piro hunters report choosing the large prime-adult animals *when presented with a choice*, as is also the case with primate troops and herds of peccaries. Thus, although the data on the ages of prey taken are imprecise, there appears to be a strong tendency to produce death assemblages biased in favor of prime-aged adults. However, some factors attenuate this pattern. If the only shot possible is a sub-adult, the hunter is likely to pursue it rather than continue a generalized search. Also, it appears that a number of non-targeted infants and juveniles are killed or captured even though they were not actively pursued independently of their mothers. Immature primates cling to their mothers, and these young individuals were often killed when their mothers were shot. Or, if their mother was killed and the infant survived, it was often easily captured and brought back to the village to be kept as a pet until it was later eaten or died from some other cause. This was the case with other species as well. A number of young capybara, peccary and tapir were captured after their mothers were killed because they hesitated to leave her body. Such a pattern could inflate the numbers of young that would appear in the archaeological records of residential sites, even though prey choice as defined by hunters active decisions is biased toward prime adults.

Sexual dimorphism appears to play a role in intra-specific prey choice. In general, dimorphism in body size should be associated with a bias in favor of the larger sex during simultaneous encounters with multiple individuals. Woolly, howler and capuchin monkeys normally are encountered in groups, and males are significantly larger than females. The data suggest the proportion of male howlers and capuchins monkeys killed by shotgun tends to be higher than would be expected if hunters were taking each sex in the proportions found in the reference populations. The samples are too small, however, to demonstrate this conclusively. Peccaries, capybara and spider monkeys do not display pronounced sexual dimorphism. Male and female peccaries and capybara are killed in more or less equal numbers, suggesting no sex-specific vulnerability for these prey species.

The cases of woolly and spider monkeys, however, suggest that behavioral aspects associated with sexual dimorphism can interact with technology in determining intra-specific prey choice. Even though male woolly monkeys are larger than females, bow hunters in Yomiwato kill more females than males. While the Machiguenga report that big males are the preferred target, bow hunters frequently miss their first shot and the troop flees. Individuals that flee more slowly and/or are lower in the forest canopy can be killed more quickly and therefore require shorter handling times. Machiguenga hunters report that pregnant females and females with offspring drop lower in the canopy and flee more slowly than males because of the burden of a fetus or a dependent infant. At least six of the 13 female woolly monkeys and 6 of the fourteen female adult spider monkeys killed in Yomiwato were either pregnant or had dependent young. Even though females provide less meat per kill, it appears that the relative inaccuracy and low kill power of bows, and associated lengthy pursuit times, result in females yielding higher energetic returns than males.

Shotgun hunters generally have the same opportunities as bow hunters to select the largest individuals from a group of monkeys, but shotgun hunters are much more likely to kill them in their first attempt. However, during the early phase of encounter, males tend to explore threats more often. This makes males especially vulnerable to the initial blast from a shotgun. An analogous situation is found with primates species captured with dart guns, a weapon nearly as effective as shotguns. J. Supriajatna (personal communication) reports that a disproportionate number of macaques darted in the Sulewasi rainforests are males.

Some Additional Implications for Archaeology

The difference between shotgun and bow technology may be abstracted in a way that is useful for modeling the difference between projectile weapons (such as atl-atls and heavy bows) and thrusting spears or other hand-held weapons simply as a function of hunters' proximity to prey and relative kill power. Shotguns allow hunters to kill animals at greater distances than bows and are much more effective per attempt in killing animals. Our results suggest that *bow hunters engage in longer pursuits, hunt in larger more cooperative groups, and may target more vulnerable animals than shotgun hunters.* It is possible that relative to projectile weapons, thrusting spears and hand-based kill techniques require longer, more difficult pursuits. If prehistoric hunters without free-flying projectile weapons were to pursue large game, we believe on the basis of our findings that they would have emphasized more cooperative kill techniques. We also expect Pleistocene hunters without projectile weapons might bias their search strategy in favor of ambush sites, such as stream crossings, that specifically disadvantage animals in close pursuits. These hypotheses can be tested by examining the location of kill sites and the age and sex composition of kill assemblages, associated with different tool technologies in prehistoric archaeological records.

ACKNOWLEDGMENTS

This research was conducted in the context of a larger project concerning the comparative ecology of diet, health and parenting among the native populations of the Manu region, and was funded by the Charles Lindbergh Foundation, the Tinker Inter-American Research Foundation, a University of New Mexico Student Resource Allocation Grant to Alvard, and a National Science Foundation grant (BNS-8717886), the L. S. B. Leakey Foundation Hunter-Gatherer Fellowship, and a Faculty Resource Allocation and Biomedical Support Grant from the University of New Mexico to Kaplan. Without the assistance of Kate Kopischke and Teslin Phillips, who participated in data collection and research design, this project would not have been possible.

NOTE

1. If the hours of men with bows that on occasion accompanied the shotgun hunters are considered, the number of hunter-hours observed in Diamante increases to 292.5 and the return rate in Diamante decreases to 0.99 kg per hour.

6

Nonselective Small Game Hunting Strategies: An Ethnoarchaeological Study of Aka Pygmy Sites

Jean Hudson

INTRODUCTION

One of the questions of major interest to those studying early hominid life is when and how hunting became an important part of subsistence. To approach this question we need to be able to recognize various types of hunting strategies in the faunal record and differentiate them from scavenging, as well as from natural processes of bone accumulation. Nonselective hunting strategies are of special interest in this regard because the resulting faunal assemblages may resemble those accumulated by certain nonhuman agencies (e.g., Stiner 1990b).

In this discussion, *nonselective* hunting simply means that the *hunters* do not choose their prey individually according to any criteria of age or gender. This definition does not take into account biases introduced by the behavior of prey. In some cases the hunt is restricted to a single species, as with bison drives. In other cases the technique is not selective for a particular taxon, as is often the case with net hunts and certain types of snares. These nonselective techniques can be contrasted with selective techniques, such as stalking or ambush hunting, in which the hunter may actively select certain, preferred individuals from a group of prey, according to age and gender (see also Alvard and Kaplan, Chapter 5 of this volume).

Another aspect of hunting strategy relevant to the evolution of cultural behavior is the size of game obtained and the degree of cooperation and type of technology required to obtain it. Some of our assumptions about whether an assemblage is the result of hunting or scavenging, and about the skills of the predator responsible, are contingent upon the size and behavior of the prey animals. The role of small game is often neglected in models of human hunting, yet these prey species supply much of the meat consumed by some hunting societies, particularly in tropical regions.

ETHNOARCHAEOLOGICAL EXAMPLES

Recent ethnoarchaeological research with the Aka pygmies, who rely heavily on net hunting and trapping small animals, provides an opportunity to study the type of archaeological record that nonselective, small game hunting leaves behind. The Aka are foragers living in the tropical forests of the Central African Republic. The data presented here concern the faunal assemblages from two modern Aka short-term *residential* sites, one of them a net hunting camp and the other a trapping camp. The analysis focuses on the age distribution of hunted prey and the body sizes of the species hunted.

Net hunting is a nonselective technique that typically involves the cooperation of over 30 people, including both men and women. An area of the forest is enclosed by linking a series of nets that stand about a meter high and driving the encircled game towards this barrier, where the animals become entangled and can be dispatched, typically with a spear, knife or by knocking the animal's head against a tree. Any animal startled into running can be caught in this way, but the species most frequently killed are duikers, small artiodactyls adapted to this thickly-forested environment. There is no deliberate selection apparent on the part of the hunters for animals of a particular age or gender.

Trapping is also a relatively nonselective technique. Typically, the trap captures the prey animal when the hunter is not present, and thus the hunter cannot select the age or gender of prey. The design and placement of the trap may serve to select prey according to other criteria, however. Some types of snares are aimed at specific species, such as forest rat, giant squirrel, or porcupine. Others are more general in purpose and are effective with a variety of duikers and small carnivores.

The majority of the game animals caught by the Aka group studied were species with an average body weight of less than 25 kg. Ninety percent of the combined total MNI for the two sites discussed came from species weighing 5 kg or less. These modern assemblages are the results of two types of nonselective hunting, one of which requires the cooperation of a large group of men and women and results in some highly structured patterns of food sharing at the residential base (Hudson 1990a, 1990b).

MORTALITY PROFILES

One of the approaches most commonly used to differentiate between faunal assemblages associated with different hunting techniques is the analysis of the age structure of the hunted population. This is typically done using teeth, assigned to different age classes, based on eruption and wear or some other measure such as annuli or crown height.

Of the two Aka sites discussed here, only one of them, the net hunting camp called Miseteke-10, had an adequate sample of ageable teeth. The species best represented was the blue duiker, *Cephalophus monticola*. Forty-one hemi-mandibles were recovered, representing a minimum of 27 individuals. Given the relatively short lifespan of the blue duiker (an ecological longevity of roughly 5 years and a maximum age of 6 to 12 years in captivity [Crandall 1965; Mentis 1972]) and the number of age intervals I used to construct the profiles (8 age classes), the sample size of 27 individuals is assumed to be adequate (see Lyman 1987b).

Complete or nearly complete hemi-mandibles were much more common in the collection than were isolated teeth, making determinations of age by eruption and wear relatively easy. This was fortunate as the only data available for this species that describe the age structure of wild populations and link known ages of individuals to their dentition is also based on tooth eruption and occlusal wear categories (Dubost 1980). For the sake of comparison with age profiles of other species, two alternative methods of age determination were also attempted, both based on crown height.

Before discussing the mortality profile generated from the archaeological assemblage, a brief review of the expectations about living age structure of mammal populations is useful (see also Chapter 1).

The living-structure (or catastrophic) mortality profile represents both the structure typical of all living mammalian populations and that of death populations caused by catastrophic events, natural or human-caused (e.g., Klein *et al.* 1983; Levine 1983). All age classes are equally vulnerable. The distribution peaks in the youngest age class, and the profile declines continuously from youngest to oldest age classes (Fig. 6.1, top). This is

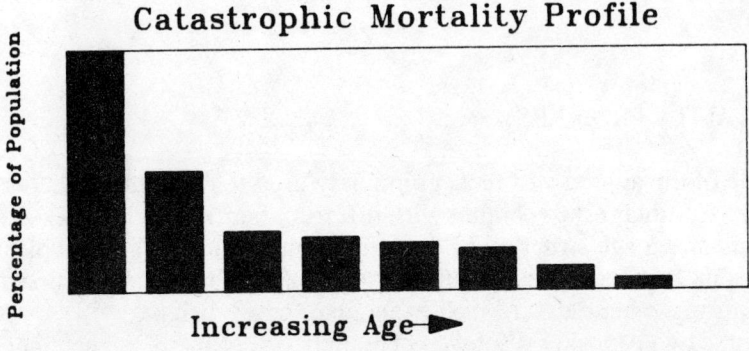

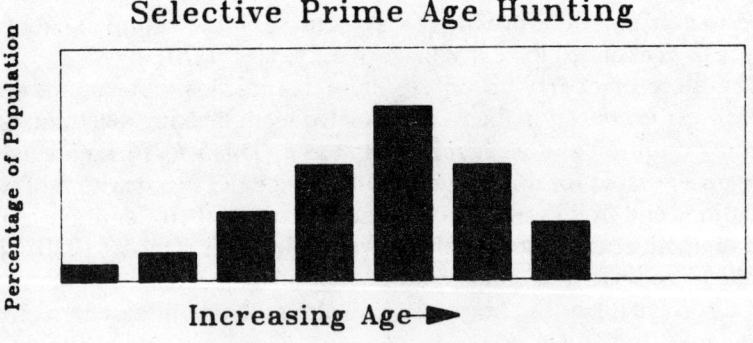

Figure 6.1. Two hypothetical mortality profile models: catastrophic (living-structure) and selective prime-age hunting (from Levine 1983; Stiner 1990b).

the type of profile usually hypothesized for nonselective hunting, because the procurers try to take whatever they encounter.

Alternatively, a common type of selective hunting of relatively large ungulate prey by humans involves focusing on prime adult prey (Stiner 1990b and Chapter 8; see also general discussion by Frison, this volume). This kind of selectively by hunters results in an age distribution that peaks in the middle of the lifespan (Fig. 6.1, bottom).

PROFILES FOR NONSELECTIVE HUNTING

The age profile based on the archaeological remains recovered from the Aka net hunting site is something of a surprise (Fig. 6.2). Given that net hunting is a nonselective technique from the human perspective, one would expect it to produce an age profile similar to that of the live population. Instead, it most closely resembles the selective prime age hunting model, although the observed hunting profile peaks in the subadult age class while the selective prime-dominant model peaks in the young adult

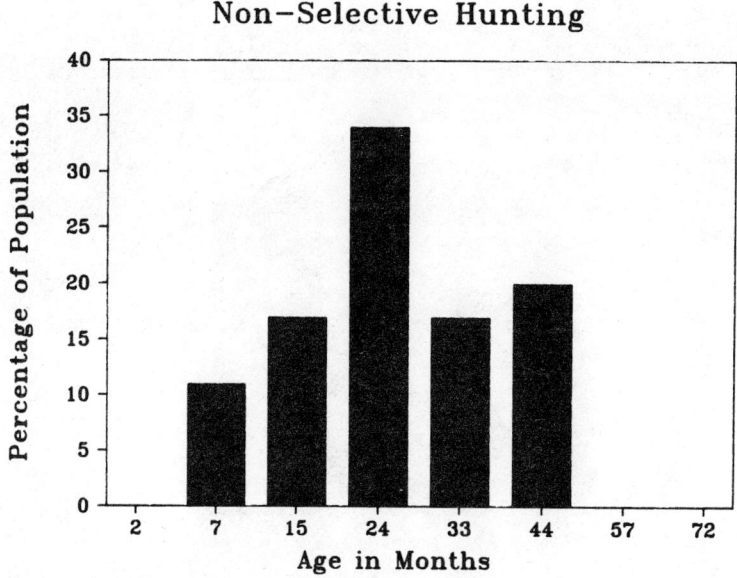

Figure 6.2. Mortality profile for the archaeological assemblage of blue duikers net-hunted by Aka foragers.

classes. This may be an important distinction, but it also must be recognized that the overall shapes of both are very similar. Most importantly, the observed hunted case fails to resemble the expected living-structure profile shown in Fig. 6.1.

Since a nonselective hunting mortality profile is expected to look like the age distribution of a live population, it is appropriate to examine what census data are available for living blue duikers, the most common prey item taken in nets (Fig. 6.3). These control data are the result of four years of live censusing of a wild population in Gabon (Dubost 1980), and the sampling technique was very similar to an Aka net hunt. Nets were used to completely surround the hectare being censused and game was driven into the net. This procedure was repeated over a 74 hectare area. It should be noted that the census covered considerably less area than is normally covered by Aka net-hunters during the course of occupation of a residential camp and differences in sample size may thus influence results.

The net hunting mortality profile resembles the censused profile with one notable deviation. While both profiles lack the high abundance of very young expected of natural populations, Dubost's live census data also

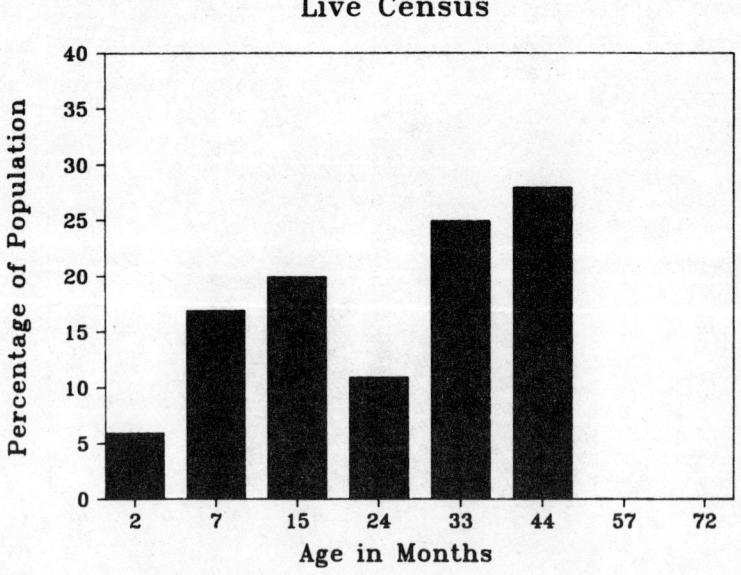

Figure 6.3. Age profile based on live census data for a wild population of blue duikers, also captured with nets.

show a relative lack of subadults in the 24 month age category. In the archaeological sample the opposite is true; 24 month subadults are more abundant than any other age class.

One possible explanation for the patterning in the profiles lies in the interaction between age-specific behaviors of the prey species and the human hunting technique. The two patterns, the lack of the very young and the discrepancy in the abundance of subadults, can be addressed separately.

The surprising similarity of the Aka nonselective net-hunting profile to the theoretical selective prime age hunting model may be the result of the failure of the particular hunting technique to capture the very young in representative numbers. Behavioral observations by Dubost (1980) suggest that blue duikers less than 10 months old are more likely to remain quiet and stay hidden than to jump and run when startled. This makes them less likely to be taken in either a net hunt or a net census and explains their absence in both profiles.

The difference between the net-hunting profile and the net-based census profile in the relative abundance of subadults may also be explained by the interaction between hunting and duiker social behavior. Adult blue duikers usually live in family groups consisting of one male, one female, and their offspring of less than a year and a half. Each family group occupies a scent-marked territory of 2.5 to 4 hectares. Subadults, in contrast, live singly and are more mobile, roaming as they seek mates and territories of their own (Dubost 1980).

When young duikers grow to be subadults in an area already well-packed with adults, they are forced to migrate out of the immediate area and into adjacent zones in order to establish their own territories. Thus, a well-packed area might be characterized by an age profile in which adults and very young animals are relatively abundant while subadults are uncommon.

If such a "packed" area is subjected to nonselective hunting, the mortality profile would resemble the age profile for Dubost's live census data. Subsequent to such hunting, formerly occupied territories would open up as a result of adult mortality. This would create a situation attractive to subadults seeking territories and the surviving local population might be supplemented by roaming subadults from adjacent areas. The age profile of this replacement population would show a peak among subadults, similar to that seen in the profile from Miseteke-10.

The hunting strategy of the Aka in the study area is nomadic as well as nonselective in terms of age. During the dry season (November-July), when they engage in net-hunting, the Aka generally move camp every two to two and a half months. Net hunts are conducted almost every day within this period, typically within a 10 to 15 km radius of the camp. Participants move in a series of loops such that each day a new area is hunted adjacent to that of the previous day. When net hunts yield insufficient game to support the group, the group relocates.

This can be expected to have the effect of seriously decimating the population in the area surrounding an occupied camp and then letting it "lay fallow" after the camp is abandoned. In a region that is heavily exploited by Aka groups or other hunters, one would expect a higher proportion of subadults. In a region under less hunting stress, one would expect the accumulation of adults with established territories.

While better-controlled census data over a larger area are needed to fully test this hypothesis, one possible interpretation of these profiles is that the mortality profile from Miseteke-10, the net hunting camp, represents nonselective hunting in an area already heavily hunted, while the live census profile represents the local natural population structure in an area that is filling up with established adults and has not been heavily hunted recently.

The results of this ethnoarchaeological test case suggest that nonselective hunting strategies must be modeled not only in terms of the theoretical distribution of a living-structure profile, but also taking into account the age-specific behavior patterns, social habits, and spatial needs of the species. If the behavior of the very young successfully protects them from the particular hunting technique, the hunting profile, even when nonselective of age in terms of the hunters' choices, may more closely resemble the selective prime-age model than the living-structure profile.

It should also be kept in mind that age profiles are dynamic and that a local pattern may represent only one variation in the larger temporal and spatial scheme. The size of the sample should be evaluated for its ability to represent the larger population, taking into account population density and social structure. The potential impacts of localized hunting stress should also be considered. In the case of relatively small animals with high population densities and family territories, such as the blue duiker, it may be possible to use mortality profiles to monitor shifts in hunting stress as they are reflected in fine-scale fluctuations in the relative abundance of adults and subadults.

METHODS FOR CONSTRUCTING MORTALITY PROFILES

The age profiles used in the above discussion are based on categories of tooth eruption and occlusal wear. Baseline data linking such categories to known ages were available and complete tooth rows were recovered archaeologically. Alternatively, mortality profiles for archaeological assemblages are often based on the measurement of crown heights, since this method can be applied to isolated teeth and to species for which known-age data are limited (Klein and Cruz-Uribe 1984). The assumption involved is that teeth will wear as the animal ages and that diminishing crown heights are correlated with increasing age in a predictable way (see also Gifford-Gonzalez, Chapter 4 of this volume).

This relationship has been modeled as both a linear and curvilinear function. The linear model assumes that the tooth will wear and the crown height diminish at a relatively constant rate. The curvilinear relationship modifies this with the assumption that the tooth will wear faster when the animal is younger and the crown is higher with less occlusal surface area, and that the wear rate will slow down as the animal becomes older and the tooth becomes flatter (Klein and Cruz-Uribe 1983, 1984; Klein *et al.* 1981, 1983; Spinage 1972, 1973).

These two crown height models have been applied to the archaeological assemblage from Miseteke-10 (Gordon n.d.) using an 8 cohort classification system. The results can be compared with those based on eruption and occlusal wear categories derived from animals of known age (Fig. 6.4). The eruption and wear categories have been adjusted to compensate for the effect of unequal age class intervals.

The relationship between the two crown height methods is as follows: the linear regression appears to over-emphasize the older age classes slightly, whereas the curvilinear relationship appears to over-emphasize the younger age classes to a more significant degree. The age profile based on eruption and occlusal wear categories falls between the two. If we accept that the eruption and occlusal wear profile is the best estimate, assuming that Dubost's assignment of real ages to particular eruption stages is accurate and that his criteria have been applied correctly to the archaeological sample, then the question arises whether the Spinage/Klein formula as it now stands might result in age profiles biased in favor of younger age classes (see also Gifford-Gonzalez, this volume).

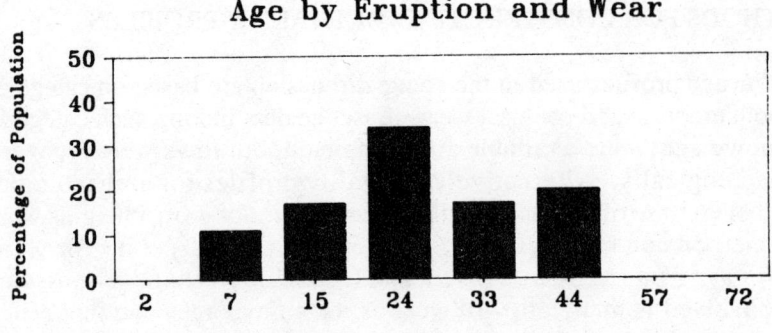

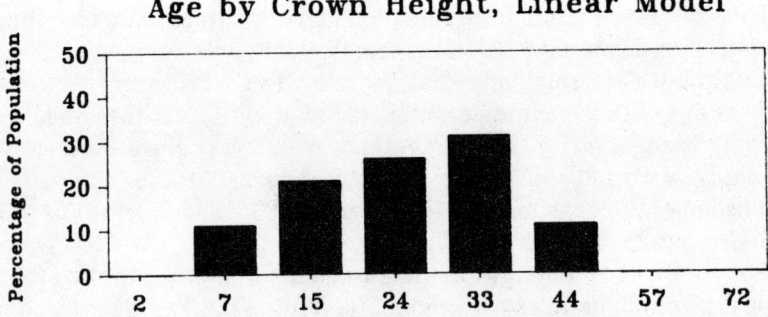

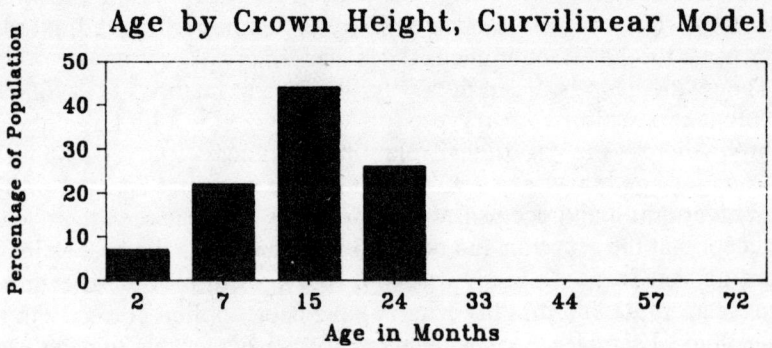

Figure 6.4. Comparison of the results of three alternative methods for constructing mortality profiles from Aka archaeological assemblages: tooth eruption and occlusal wear (top), crown height based on a linear wear model (middle), and crown height based on a curvilinear wear model (bottom).

GAME SIZE AND HUNTING TECHNIQUE

A second major issue of interest in modeling the ways humans obtain meat concerns game body size and its association with particular hunting techniques. Ethnoarchaeological data on the Aka provide an opportunity to add to our models. Most of the hunting techniques used by the Aka are best suited to the capture of small species weighing less than 25 kg in thickly-forested habitats (see also Alvard and Kaplan, Chapter 5).

The two hunting techniques most commonly used, net hunting and trapping, are associated with different degrees of cooperation and different taxonomic emphases. In both cases the equipment used is perishable, consisting of cordage, vine and/or wood, and is unlikely to survive archaeologically. Thus, the composition of the faunal assemblage represents one of the few types of clues to use of these techniques in the past.

Trapping among the Aka typically involves only one or two hunters and is especially well-suited to the capture of rodents. While rodents are generally too small to be caught efficiently by other techniques, trapping requires little investment on the part of the hunter. A line of traps can be set and checked in a few hours and can provide a reliable source of protein for a small group of people.

Net hunting requires the cooperation of over 30 adults and is especially well-suited to the capture of small and medium-sized duikers. *Adult* duikers, like many artiodactyls, will flee when startled. Their long-legged build allows them to cover distances rapidly but makes them relatively easy for a hunter to track visually in a heavily-vegetated environment, at least in contrast to short-legged rodents and carnivores. When these duikers encounter the net, their relatively small body size prevents them from immediately breaking through and makes it possible for a hunter to pin them.

The contrast between the two techniques in taxonomic emphasis is also apparent in the archaeological record accumulated at the two Aka sites, Miseteke-10 (a net hunting camp) and Mokumbokumbo (a trapping camp). This contrast is illustrated in Fig. 6.5 (also Table 6.1). An understanding of the behavioral differences between the taxa most frequently caught and the strengths of particular hunting techniques provides clues to the dominant hunting strategy used while occupying a particular residential site.

Combining the taxa by size class demonstrates the similar emphases on small game common to both sites in spite of other differences (Fig. 6.6). Three size categories have been used here: small game of 5 kg or less (e.g., blue duiker, brush-tailed porcupine); medium game between 5

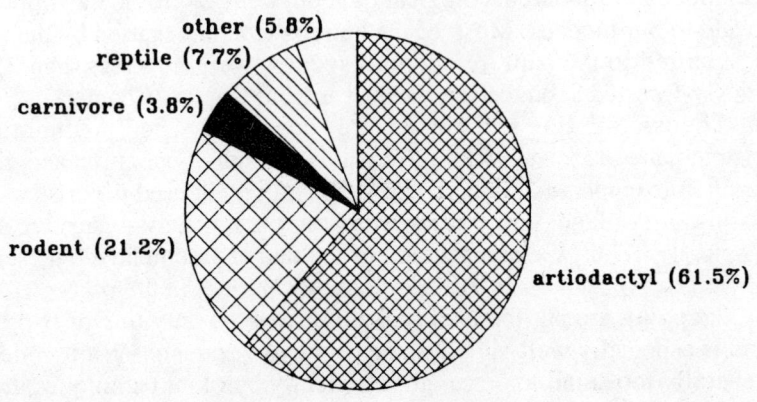

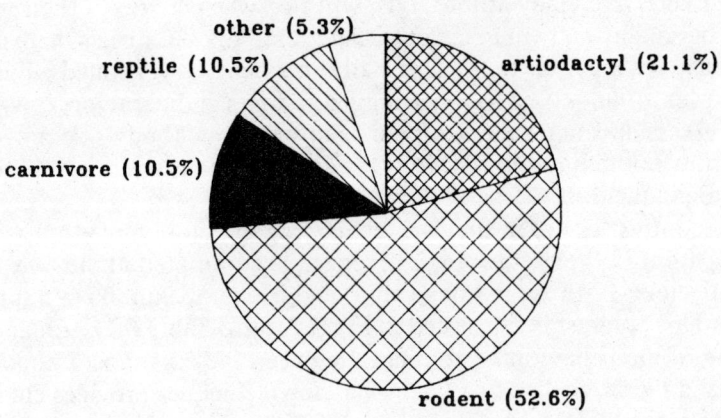

Figure 6.5. Taxonomic composition of hunted game remains at two modern Aka camps, based on MNI values for recovered bone.

Table 6.1. List of species recovered archaeologically from two modern Aka camps.

species	live weight (kg)	Miseteke 10 (net hunting)		Mokumbokumbo (trapping)	
		MNI	NISP	MNI	NISP
artiodactyls:					
Cephalophus monticola (blue duiker)	5.0	27	258	4	51
Cephalophus callipygus (Peter's duiker)	20.0				
Cephalophus dorsalis (bay duiker)	22.0	4	118	3	67
Cephalophus leucogaster (Gabon duiker)	13.0				
Cephalophus silvicultor (yellow-backed duiker)	60.0			1	26
Hyemoschus aquaticus (chevrotain)	13.3	1	1		
rodents:					
Atherurus africanus (brush-tailed porcupine)	3.0	7	56	7	77
Cricetomys emini (forest rat)	1.0	2	9	9	106
Funisciurus pyrrhopus (red-footed squirrel)	.25	2	3	3	5
Protoxerus strangeri (giant squirrel)	.6			1	2
carnivores:					
Nandinia binotata (two-spotted palm civet)	3.35			2	7
Genetta sp. (genet)	2.25	1	1		
Bdeogale nigripes (black-legged mongoose)	3.2			1	3
Atilax paludinosus (marsh mongoose)	3.2	1	1		
Herpestes naso (long-snouted mongoose)	3.2			1	1

(continued)

Table 6.1. (Continued)

species	live weight (kg)	Miseteke 10 (net hunting)		Mokumbokumbo (trapping)	
		MNI	NISP	MNI	NISP
primates:					
Cercopithecus sp.	4.5	1	4	1	3
(Cercopithecid monkeys)					
pangolins:					
Manis sp.	2.25	1	7	1	4
(pangolins)					
reptiles:					
Kinixys erosa	1.5	3	38	2	46
(land tortoise)					
Python sp.	~8.0	1	131	1	5
(pythons)					
Varanus niloticus	~10.0			1	7
(monitor lizard)					
birds:					
Aves	?	1	7		
undifferentiated bird					

kg and 25 kg (e.g., bay duiker, Peter's duiker); and large game of 25 kg or more (e.g., yellow-backed duiker). These size categories are scaled to the present collection. Even the largest of the species represented, the yellow-backed duiker weighing roughly 60 kg, would be considered relatively small in comparison to many of the taxa commonly recovered from certain modern and prehistoric hunter-gatherer sites elsewhere in the world.

Modern Aka hunting strategies and the archaeological consequences in residential sites thus represent a useful addition to our understanding of the relationship between game body size and socioterritorial behavior and the hunting techniques used by humans. The Aka provide an example of the viability of small game hunting, demonstrate the potential importance of rodents and the role that trapping can play in subsistence strategies, and

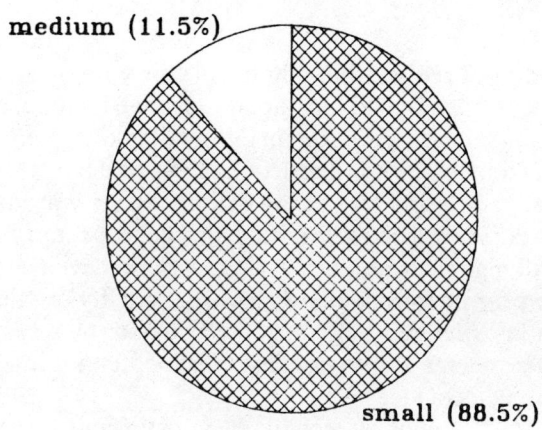

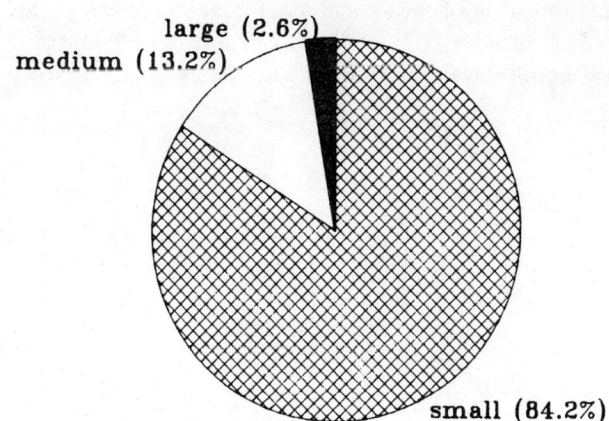

Figure 6.6. Body size classes of hunted game remains at two modern Aka camps, based on MNI values for recovered bone.

show that small game hunting can be associated with cooperative techniques that involve large groups of men and women.

CONCLUSIONS

In concluding, I return to the theme of this volume and summarize what this ethnoarchaeological look at nonselective hunting, as defined by the behavior of hunters, adds to our understanding of age profiles of death assemblages and their ability to reflect hunting strategy. Our assumption that nonselective hunting strategies will necessarily produce a living-structure mortality profile may be too simple, especially in situations involving small highly territorial forest species. To develop more comprehensive models for interpreting nonselective mortality profiles we need to incorporate a knowledge of age-specific behavior patterns that characterize the prey species and to consider their potential interaction with particular hunting techniques.

A wider range of hunting techniques are available to human hunters than is sometimes appreciated in our modeling of the evolution of human foraging and social behaviors. Some of these techniques, such as driving small game towards nets and trapping, make use of equipment that has poor potential for surviving archaeologically. Other lines of evidence, such as the composition of the faunal assemblage in terms of the size, age and behavioral characteristics of the taxa represented, should be incorporated into our models and analyses about prehistoric strategies for obtaining meat.

ACKNOWLEDGMENTS

This research was funded by the L. S. B. Leakey Foundation and the University of California at Santa Barbara, and was conducted with the kind permission of the Ministére de la Recherche Scientifique et Technique of the Central African Republic.

7

Prey Size and Age Models of Prehistoric Hominid Scavenging: Test Cases from the Serengeti

Robert J. Blumenschine

INTRODUCTION

The hunting and scavenging debate in paleoanthropology is important for resolving more fundamental issues in human evolution. The size range of animals that hominids hunted, the sources of carcasses and dominant tissue types acquired by scavenging, and the seasons and habitats in which carcass acquisition occurred and its regularity all are important to interpretations of hominid technology, ranging patterns and sociality based on early archaeological sites.

Indirect evidence from the predatory and scavenging strategies of modern carnivores and primates, though suggestive, cannot resolve the debate alone. The contradictory results of numerous attempts at argument based solely on analogy (e.g., Butynski 1982; Hasagewa *et al.* 1983; King 1975; Schaller and Lowther 1969; Strum 1983; Teleki 1981) reveals only the broad limits of the methods hominids may have used to procure carcasses. Unfortunately, these limits could also disqualify any hominid foraging strategy no longer in existence. For example, the mixed arboreal-terrestrial scavenging opportunities provided today by leopard kills cached in trees and lion kills abandoned on the floor of riparian woodlands in the Serengeti may have supported hominid carnivory in the past (Cavallo and Blumenschine 1989), despite the fact that no modern carnivore occupies this niche. Likewise, Schaller and Lowther (1969) have suggested that hominids once may have occupied the currently open niche in African savanna-woodlands for a diurnal social predator and scavenger. Dismissal of scavenging by hominids based on the extremely rare scavenging of the most predatory nonhuman primates (e.g., Butynski 1982; Strum 1983; Teleki 1981; Tooby and DeVore 1987) ignores the very possibility of unique, large-carcass, tool-mediated carnivory on the part of early hominids. Similarly, the general rule that most larger carnivores both hunt and scavenge does not by itself prove such behavioral flexibility among omnivorous hominids as some have argued (e.g., Potts 1988b). Reliance on this analogy, which is based on trophic similarity, ignores the fact that hominids do not share the physiological and anatomical commitment to meat-eating typical of the Carnivora (e.g., Blumenschine 1987, 1988a). Hominids may well have supplemented a plant food diet with mixed hunting-scavenging, but recourse to modern behavioral analogs alone will never demonstrate the existence of such versatility or whether it was based on carcass size, season, habitat, or some other niche dimension nonexistent or as yet undetected in the modern world.

Three lines of fossil evidence are available for determining the source and completeness of carcasses acquired by hominids. All of these derive from the zooarchaeological record, including (1) skeletal part profiles of archaeological bone assemblages (e.g., Binford 1981, 1984; Blumenschine 1986a; Bunn and Kroll 1986; Potts 1983), (2) stone tool butchery patterns either alone or in relation to evidence for bone modification by carnivores (e.g., Binford 1978, 1984; Blumenschine 1988b; Bunn and Blumenschine 1987; Bunn and Kroll 1986; Shipman 1986a), and (3) age and size profiles (e.g., Vrba 1975; Klein 1982a). Together, the three approaches have the potential for providing partially independent diagnoses of hominid subsistence patterns.

The power of each line of zooarchaeological evidence depends on the development and appropriate application of specific taphonomic models derived from experimental and naturalistic observations in the modern world. The "middle range" approach to the hunting and scavenging debate, as developed by Binford (1981), demands the interpretation of patterns in zooarchaeological data by recourse to observations of modern processes that produce similar patterns. Additionally, one must determine the mechanism(s) affecting the modern process-pattern linkage and specify the context within which the process occurs in order to justify claims of temporal uniformity and minimize problems of equifinality (Blumenschine 1988a; Gould and Watson 1982). All of these criteria for developing taphonomic models require that actualistic studies be designed specifically in response to the archaeological problem and databases at hand. Skeletal part data are currently most widely used in interpretations of the mode of animal food procurement by hominids. Patterns of skeletal representation are expected to typify scavenged versus hunted assemblages (e.g., Binford 1984; Blumenschine 1986a; Potts 1983). Indeed, some uniformitarianistic mechanisms (bone durability, nutritional yield associated with particular bones, disarticulation sequences) that potentially pattern bone assemblages have been specified, as have some of the ecological contexts in which scavenging opportunities can arise. This is not to say that a consensus for distinguishing hunting and scavenging on the basis of skeletal part profiles of archaeological assemblages exists; there is still debate, for example, over which pattern(s) and mechanism(s) are most relevant. Additional complications arise from the possible selective transport to, destruction at and/or deletion from a fossil locality of bones by biotic and physical agents (see, for example, Klein 1989; Turner 1989 for the affect of pre- and post-depositional fragmentation and recovery biases on skeletal part profiles at Klasies River Mouth).

Experimental and ethnographic observations on carcass processing and associated bone marking also have considerable potential for demonstrating the edible tissue completeness and condition of carcasses acquired by hominids. For instance, Bunn and Kroll (1986, 1988) and Bunn (1986) have argued that the presence of defleshing cut marks on the midshafts of "meaty" limb bones preserved at several Early Pleistocene sites is evidence that hominids had early access to carcasses, possibly through hunting. However, the relevant actualistic observations that could effectively distinguish defleshing of whole muscle masses from removal of scraps, as would be expected from scavenging (Blumenschine 1986b), have not been conducted. Binford (1984) argues that the presence of hack marks on some bones in the Klasies River Mouth fauna indicates processing of partially mummified or otherwise stiffened, scavenged carcasses, an inference based on ethnographic observations of Nunamiut use of frozen carcasses (Binford 1978). Because butchery-marking within a given technological context is constrained by the anatomy of carcasses (which for present purposes has not changed during the Quaternary), one should be able to link particular locations and frequencies of butchery marks to the completeness and condition of carcasses at the time of procurement, and hence hominids' means of procurement. That this is currently not possible is signaled by Shipman's (1986a, 1986b) controversial use of the location of cutmarks on long bones to diagnose scavenging by Early Pleistocene hominids at Olduvai Gorge (Bunn and Blumenschine 1987; Gifford-Gonzalez 1989; Lyman 1987a).

Given the potential problems with skeletal part profiles and butchery patterns, size and age profiles of archaeological faunas provide a potentially important, independent line of evidence for diagnosing the mode of carcass procurement by hominids. Indeed, Vrba's (1975) use of size and age profiles represents one of the first attempts to address this aspect of hominid subsistence on the basis of fossil evidence. Klein (e.g., 1978, 1981a, 1981b, 1982a, 1987) has also used data on age mortality profiles of paleontological and archaeological ungulate assemblages from South Africa.

Nonetheless, the basic premises of size and age profiles as criteria for distinguishing hunting and scavenging in prehistoric assemblages have not been tested or demonstrated fully against a body of actualistic data collected with this purpose in mind. Klein (1978) has used modern African buffalo mortality profiles, including those resulting from lion predation, to interpret the age structure of Cape buffalo in archaeological contexts. More recently, Lyman (1987b) has demonstrated that modern sport-hunting pressure affects mortality profiles of cervids killed catastrophically during the 1980 eruption of Mount St. Helens. More studies such as these are needed to lend important baselines for interpreting prehistoric mortality profiles. In particular, we must establish diagnostic mortality patterns and the mechanisms that produce them, as well as assess the extent to which changing contexts of carcass procurement by hominids will affect the size and age profiles of scavenged and hunted assemblages.

Here, I provide one such test of mortality profile and carcass size models of scavenged assemblages. The test is based on the size and age distribution of 235 fresh carcasses resulting from predator kills and, in fewer instances, natural deaths in the Seronera woodlands and southern plains areas of Serengeti National Park, Tanzania. All carcasses were located during the day by searching for animals already dead, rather than by following predators in anticipation of a kill (see Blumenschine 1986b, 1987 for details of the study site and methods). As such, the Serengeti carcass sample may be thought of as representing those potentially encountered by an opportunistic diurnal scavenger that exploited the remains of both predator kills and natural deaths. By stratifying the carcass sample according to the context within which carcasses were located, I will evaluate the extent to which changes in the context of scavenging opportunities modify the sizes and ages of carcasses potentially procured. The contexts to be examined vary along several dimensions including (1) habitat, (2) season, (3) cause of death, (4) the identity of carcass consumers, particularly presence/absence of lions and spotted hyenas, and (5) the tissue completeness (yield) of the carcass upon discovery. These contextual variables are important in determining both the profitability of scavenging (Blumenschine 1987) and the projected archaeological signature of scavenging, based on skeletal part profiles (Blumenschine 1986a).

TAPHONOMIC BIASES RELATED TO CARCASS SIZE AND AGE

Corresponding bones from animals of different size and age vary with respect to durability and nutritional value (Binford 1978; Binford and Bertram 1977; Blumenschine and Caro 1986; Brain 1981; Lyman 1984; Metcalfe and Jones 1988). Taphonomic agents accordingly bias the size and age representation of carcasses in modern setting in ways that are fairly well understood and therefore potentially correctable (e.g., Behrensmeyer et al. 1979; Gifford 1981).

With respect to scavenging, carcass size has a predictable effect on the persistence of carcasses as scavengeable food resources, and therefore the chance that a carcass will be available to a scavenger (Blumenschine 1986b and Fig. 7.1). There is a positive linear relationship between the log of live carcass weight and log carcass persistence (i.e., the duration over which a carcass provides a food source). Hence, small carcasses, including the young of larger species, are less likely to survive initial consumption by predators as a scavengeable resource.

This relationship manifests itself clearly in differences between the size structure of live ungulate populations in the Serengeti (Frame 1986) and that observed in my total carcass sample (Table 7.1 and Fig. 7.2). In particular, the carcass sample shows an under-representation of size 1 and to a lesser extent size 2 individuals (see also Behrensmeyer et al. 1979). The magnitude of the difference can be evaluated using the Chi-square "goodness of fit" test. Here, the proportion of animals in each size category of the live community is expressed as a percentage of the total carcass sample (N=235). When size groups 5 and 6 are lumped to avoid obtaining invalid expected cell frequencies, a significant difference is found (Chi-square=24.5, df=4, p=0.0001).

SIZE PROFILES

Vrba (1975) proposed that the variance in the body weight distribution of fossil faunal assemblages should be sensitive to the mode of carcass acquisition. "Primary" or hunted assemblages should show a peak in a

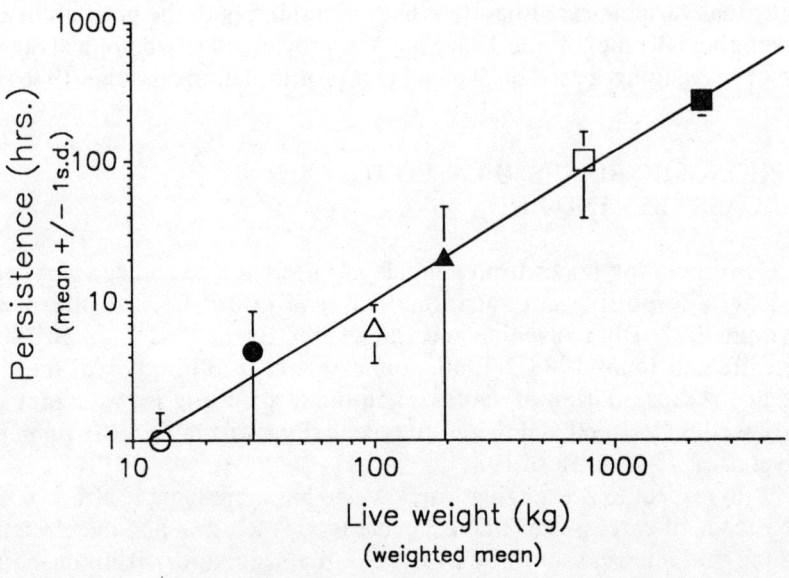

Figure 7.1. The relationship between live weight of mammalian herbivores and their persistence after death as sources of scavengeable food in the Serengeti (after Blumenschine 1986b). Live weights are weighted means for the carcasses making up each group. Open circles = small subadults (3 Thomson's gazelle subadults, 1 Grant's gazelle sub-juvenile and 2 wildebeest sub-juveniles); closed circles=small adults (3 Thomson's gazelle and 2 Grant's gazelle); open triangle = size 3 juveniles (8 wildebeest, 2 zebra); closed triangle = size 3 adults (60 wildebeest, 11 zebra); open square = large adults (7 buffalo); closed square=very large adults (2 giraffe, 1 elephant). When expressed as a log-log relationship, carcass persistence is highly correlated to live weight ($r=0.99$, $p<0.001$).

restricted body weight range, accompanied by a relatively narrow range of variation if a single predatory pattern predominated in the accumulation of the assemblage. "Secondary", or scavenged assemblages, on the other hand, would show a size distribution with larger variance. Nonselective choice and accumulation of carcasses by scavengers would ideally produce a size profile that approaches live distributions. In conjunction with

Table 7.1. Size profiles for the Serengeti carcass sample.

		\multicolumn{7}{c}{BODY SIZE GROUP[a]}	
		1	2	3	4	5	6	TOTAL
SERENGETI LIVE[b]	n	650,000	307,250	1,842,145	103,180	28,270	3,850	2,934,695
	%	22.1	10.5	62.8	3.5	1.0	0.1	
CARCASS SAMPLE								
TOTAL	n	22	10	193	7	2	1	235
	%	9.4	4.3	82.1	3.0	0.9	0.4	
lions and/or hyenas:								
present	n	5	3	122	3	0	0	133
	%	3.8	2.3	91.7	2.3	0	0	
absent	n	17	7	71	4	2	1	102
	%	16.7	6.9	69.6	3.9	2.0	1.0	
cause of death:								
natural[c]	n	1	0	30	0	0	1	32
	%	3.1	0	93.8	0	0	3.1	
predator	n	19	3	112	3	0	0	137
	%	13.9	2.2	81.8	2.2	0	0	

(continued)

Table 7.1. (Continued)

habitat:		1	2	3	BODY SIZE GROUP[a] 4	5	6	TOTAL
riparian woodland	n	1	2	48	1	1	1	54
	%	1.9	3.7	88.9	1.9	1.9	1.9	
plains	n	19	4	126	1	0	0	150
	%	12.7	2.7	84.0	0.7	0	0	
acacia woodland	n	2	4	19	5	1	0	31
	%	6.5	12.9	61.3	16.1	3.2	0	

[a] Size 1 includes Thomson's gazelle; Size 2, Grant's gazelle and warthog; Size 3, wildebeest, zebra, topi and waterbuck; Size 4, buffalo; Size 5, giraffe; and Size 6, elephant.
[b] Data on live numbers of Serengeti herbivores are from Frame (1986), based on 1977 conditions.
[c] Natural (non-predated) deaths were distinguished from predator kills if the carcass was complete upon discovery with no mammalian carnivores in attendance, if the animal had drowned, or if only vultures had fed. Such criteria could not be applied to 66 carcasses, which are excluded from this stratification of the carcass sample.

Prey Size and Age Models of Scavenging, Serengeti 129

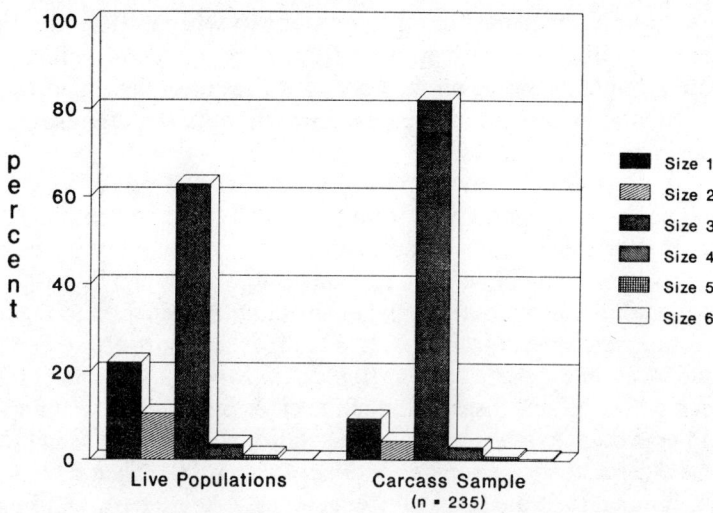

Figure 7.2. Larger mammalian herbivore size distributions of the live community and carcass sample from the Serengeti. Live numbers are from Frame (1986), based on 1977 conditions. Data and definitions of size categories are found in Table 7.1. The carcass sample, representing individuals discovered during daytime surveys, demonstrates that the quicker and more complete consumption of small (size 1 and 2) carcasses will bias the size range of carcasses encountered by a diurnal hominid scavenger from that available on the hoof. When size groups 5 and 6 are lumped (see text), the carcass sample shows a significant under-representation of size 1 and 2 animals (*Chi*-square=24.5, df=4, p=0.0001).

data on fossil fauna age profiles, Vrba used size profiles to evaluate the mode of accumulation of South African Australopithecine cave assemblages. Isaac and Crader (1981) made a similar attempt using East African Plio-Pleistocene archaeological assemblages.

Subsequently, Vrba (1980) acknowledged less confidence in the sensitivity of size profiles for distinguishing hunted from scavenged assemblages. For variance in the size distribution of faunal assemblages to be most useful, one would require currently unobtainable information on the relative species abundance of prehistoric herbivore communities from which the predated or scavenged sample is drawn. Further, Vrba cited inadequate information on the extent to which all modern predators actually killed a significantly restricted proportion of animal sizes compared to that available on the hoof. As a result of these uncertainties, size profiles have not since been used in attempts to diagnose hunted from scavenged assemblages.

A further uncertainty associated with the use of size profiles to diagnose scavenging arises from the more rapid and complete destruction of smaller size classes by predators prior to discovery of the carcass by scavengers. This factor was marshaled to explain the lower size variance of the whole of my Serengeti carcass sample compared to that of the live populations shown in Fig. 7.2. That it is an appropriate explanation is evidenced in Fig. 7.3a (data in Table 7.1), where a significant difference is seen when the size distribution of the carcass sample is stratified according to the presence or absence of lions and/or spotted hyenas upon discovery of the carcasses (Chi-square=19.3, df=2, p=0.0001 when sizes 1-2, and 4-6 are lumped to avoid low cell frequencies). These carnivores are singled-out because of their ability to consume all edible tissues on carcasses; while lions are limited in this capacity to size 1 and size 2 carcasses (including subadults of some larger size classes), spotted hyenas can deprive scavengers of any feeding opportunity on carcasses as large as size 4 adults. Accordingly, size profiles of carcasses fed upon by large carnivores show a significantly lower proportion of small (size 1 and 2) carcasses than those discovered with no larger carnivore in attendance.

These data show that the same taphonomic factors that Vrba (1975) used to associate scavenging with a low subadult:adult ratio (see below) also operate to reduce scavengers' access to carcasses of small species. In other words, if scavenging, according to Vrba, is signaled by low age variance in an assemblage, it should also be signaled by a size variance lower than that represented by the live community, and one that approaches the size distributions theoretically expected for predator accumulations.

It is not possible to assess variation among the full array of body size categories on the basis of my data. However, examination of Table 7.1 and Figs. 7.2 and 7.3a-c does show an overabundance of size 3 carcasses, and a marked under-representation of size 1 and size 2 individuals compared to their representation in the live community. When the poorly represented size groups 4 through 6 are excluded, and size groups 1 and 2 are lumped, 2 x 2 contingency tables can be constructed that relate the abundance of small (size 1 and 2) and medium-sized (size 3) carcasses in each context to live abundances for the Serengeti as a whole. Chi-square tests show that only those carcasses unattended by lions and/or hyenas

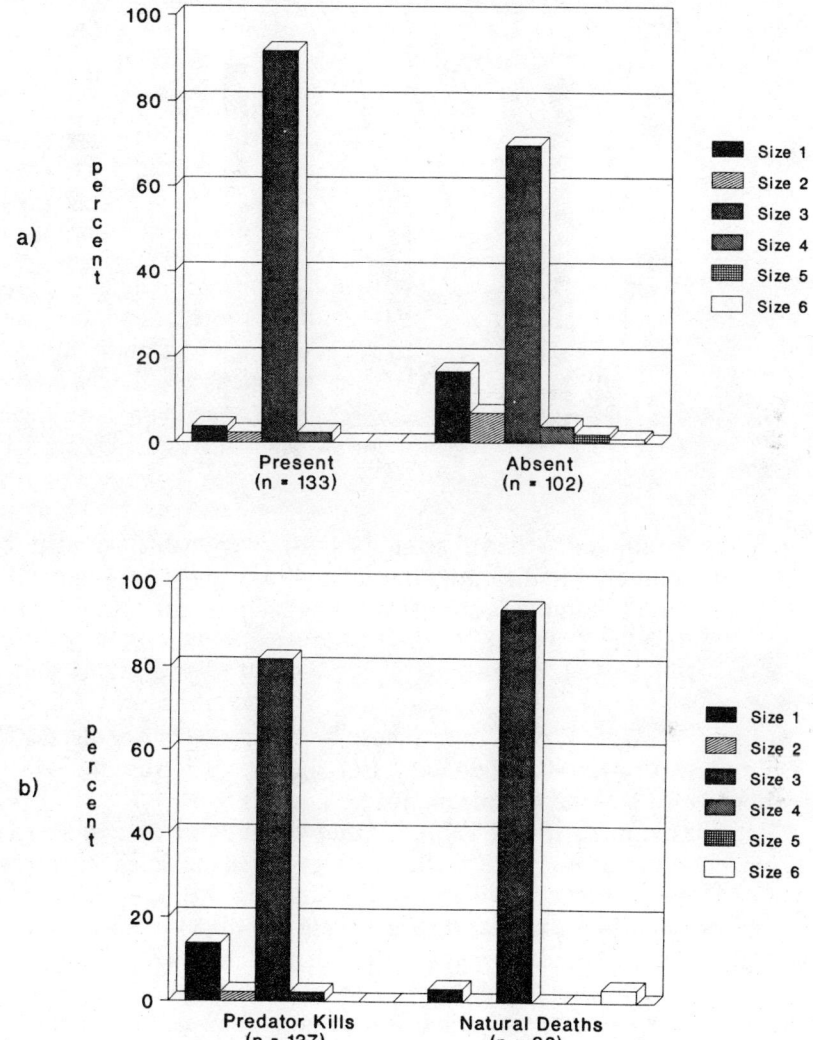

Figure 7.3. Size distributions of scavengeable carcasses encountered in different diurnal contexts in the Serengeti. The carcass sample is stratified according to (a) whether lions and/or hyenas were present or absent upon my discovery of the carcass and (b) whether the carcass resulted from predation of natural death. Raw data are found in Table 7.1.

c)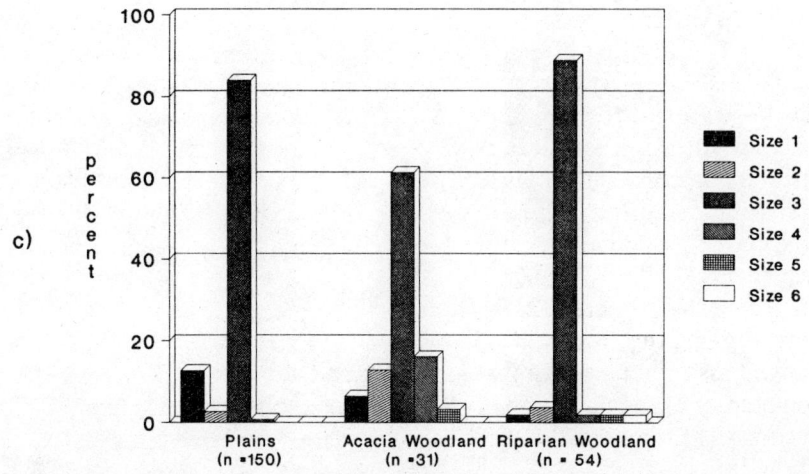

Figure 7.3 (continued). Size distributions of scavengeable carcasses encountered in different diurnal contexts in the Serengeti. The carcass sample is stratified (c) by habitat type. Raw data are found in Table 7.1.

(*Chi*-square=1.4, df=1, *p*=0.23) and those found in *Acacia* woodland (*Chi*-square=0.27, df=1, *p*=0.60) do not differ significantly from the relative live abundance of small and medium-sized animals.[1]

Further examination of Table 7.1 shows that size distributions of scavengeable carcasses are affected at a finer scale in the Serengeti, according to habitat differences and predator habits (Fig. 7.3c). For example, the Serengeti buffalo's preference for *Acacia* woodland is responsible for the high proportion of size 4 carcasses in this habitat. The lion's tendency to take size 3 prey is responsible for the very high proportion of this size class in riparian woodlands. Finally, predation by cheetah, which prefer plains habitats, is responsible for the relatively high proportion of size 1 carcasses in the plains sub-sample. Cheetah kills similarly account for the high proportion of size 1 carcasses in the predator kill sub-sample (Fig. 7.3b).

The extent of variation in the size distribution of a scavengeable sample depends on the context in which scavenging occurs, as defined by the habitat-specific composition of carnivore and herbivore species. In this light, it is appropriate to note the recent recognition of leopards as a potentially significant provider of scavengeable small (size 1 and 2) carcasses in riparian woodlands (Cavallo n.d., Cavallo and Blumenschine 1989). The addition of leopard kills to those provided by lions in woodlands would further alter the size distribution of scavengeable carcasses in this habitat, and these new data may eventually allow us to address the issue of variation raised here more fully. This highlights an important distinction between the overall carcass sizes available in an ecosystem (or in particular settings) and the subsets of the potential array of scavengeable carcasses actually exploited as constrained by the adaptations of the scavenger.

Regardless, context-specific carcass size distributions demonstrated here suggest that a single range of carcass sizes might not typify scavenged assemblages, even when these are drawn from the same general community of mammals. As realized by Vrba (1980), scavenging will not necessarily produce a broad range of carcass sizes when compared to that of the live community. The extent to which scavengeable size distributions approach and overlap those expected from specialized predation will be affected further by the extent to which a scavenger specializes on a particular source of carcasses.

For size distributions to be useful in diagnosing scavenging in the archaeological record, a number of factors need to be assessed:

1) Live size distributions in prehistoric communities (Vrba 1975, 1980), and particularly the size distributions of herbivores in particular settings, as defined by habitat and probably also by season.
2) Prey size preferences of prehistoric predator species (Vrba 1980).
3) Extent to which particular predators or initial carcass consumers differentially destroy or consume all edible tissues on smaller carcasses, therefore removing these from the total pool of potentially scavengeable carcasses.
4) More information about the adaptations of the prehistoric scavenger of interest as these might relate to carcass procurement (i.e., the realized scavenger niche).

AGE PROFILES

Age profiles have been used in two ways in attempts to distinguish hunting and scavenging in the fossil record. Vrba (1975, 1980) relied upon the preferential destruction of juveniles (presumably meant to include sub-juveniles) by predators, suggesting that "secondary" (scavenged) assemblages should have low juvenile:adult ratios compared to "primary" or hunted assemblages. Klein (1982a, 1987) considers this fundamental taphonomic principle of preferential destruction of small (including young) carcasses, within the framework of mortality. Specifically, population biologists distinguish two basic mortality profiles for stable, populations (see Chapter 1 of this volume). "Catastrophic" (also called living-structure) age profiles result from simultaneous and unselective deaths of individuals, producing carcass age profiles generally analogous to the live population age structure (e.g., Fig. 7.4, after Klein 1982a). The other mortality pattern is often termed as "attritional" (also called U-shaped) and reflects "normal" mortality (e.g., from predation, endemic disease or old age). Attritional profiles are therefore distinguished from catastrophic profiles in having a relatively high representation of old individuals, and a low representation of prime-age adults. In a 10-cohort scheme, this corresponds to roughly 10% through 40-50% of the total potential lifespan (model used for this study is taken directly from Klein 1982a, see Fig. 7.4).

Klein has used tooth crown height measurements to analyze the age structure of ungulate species, particularly small and large bovids, in a number of Middle and Later Stone Age archaeological sites and in some paleontological sites in South Africa (e.g., Klein 1978, 1981a, 1981b). Among larger bovids, the eland and the bastard hartebeest cases display catastrophic profiles in the archaeological assemblages, while those for other species for which adequate sample sizes are available (giant buffalo, Cape buffalo, blue antelope, roan antelope) are attritional (Klein 1978, 1982a).

Klein (1982) proposes that catastrophic and attritional profiles in archaeological assemblages could be produced by either hunting or scavenging by hominids. However, he interprets both catastrophic and attritional profiles for larger bovids from the South African MSA and LSA sites to reflect hominid hunting, albeit different hunting strategies (Klein 1978, 1979).

Klein prefers to equate catastrophic profiles in archaeological assemblages with hunting by hominids for two reasons. First, he (1982a, and elsewhere) cites the unlikelihood that scavenging hominids would encounter a subpopulation destroyed catastrophically before nonhuman predators had consumed the younger, less durable individuals. He also cites the low likelihood that repeated natural catastrophes occurred over the long durations that the MSA and LSA deposits he analyzed had accumulated. Rather, he argues that catastrophic profiles for larger bovids (eland and bastard wildebeest) in archaeological contexts would

135

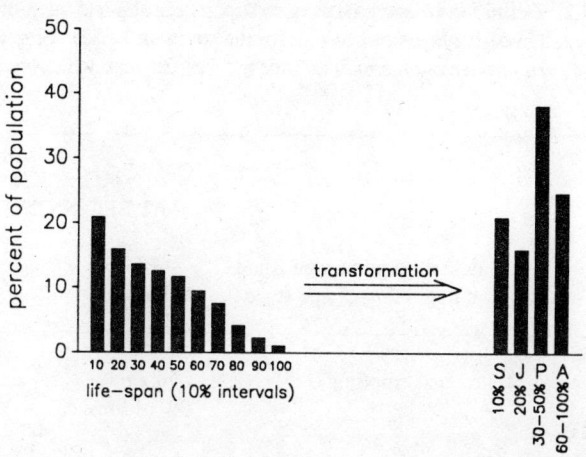

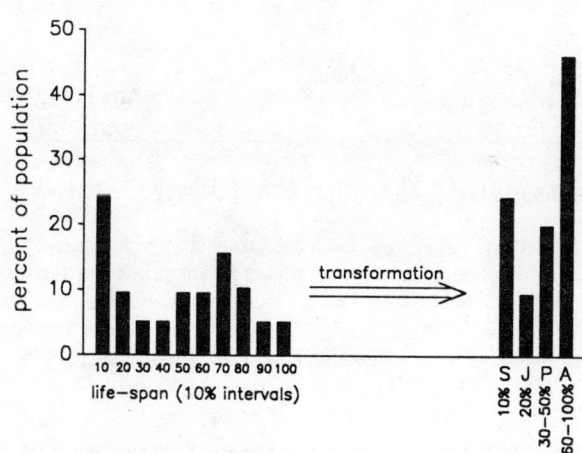

Figure 7.4. Ideal catastrophic and attritional age profiles based upon 10% increments of life-span, and their transformation into the four stage classification used here. S = sub-juvenile, J = juvenile, P = prime adult, A = aged adult. The transformed age profiles preserve the fundamental distinction of reversed proportions of prime and aged adults between catastrophic and attritional mortality. See Table 7.2 for states of dental eruption and wear corresponding to the four age categories.

Table 7.2. Definition of age groups based upon eruption and wear of mandibular teeth, and equated to percentages of lifespan for the southern African blue wildebeest (*Connochaetes taurinus taurinus*) and for the East African white-bearded wildebeest (*C. t. hecti*).[a]

		AGE AND % LIFESPAN AT BEGINNING OF AGE CLASS	
age class	dental eruption-wear state at beginning of age stage	blue wildebeest	white-bearded wildebeest
sub-juvenile	p2 just erupting	0 months 0% of lifespan	0 months 0% of lifespan
juvenile	deciduous premolars worn, m2 erupting	15 months 6.9% of lifespan	20.5 months 9.5% of lifespan
prime adult	all permanent premolars and molars in wear, all infundibuli present	38 months 17.6% of lifespan	47 months (est.) 21.9% of lifespan
aged adult	one infundibulum of m1 gone	96 months 44.4% of lifespan	120 months (est.) 55.6% of lifespan

[a] Sources are Atwell (1980) and Talbot and Talbot (1963), respectively.

Note: Ages at the beginning of each age group are given; because Talbot and Talbot's data were incomplete, these were estimated for the prime and aged adult classes of the white-bearded wildebeest by assuming the same delay in their maturation relative to the blue wildebeest seen for the sub-juvenile and juvenile stages. Percentage of lifespan estimates assume an 18-year potential, or "ecological," longevity (Klein 1978).

more likely result from driving of whole herds over cliffs or into natural traps at Klasies River Mouth and Nelson's Bay Cave, respectively (see also general discussion of driving prey by Frison, Chapter 2 of this volume). Those smaller bovids showing catastrophic age profiles and attributed to hominid hunting by Klein are suggested to have been trapped or snared, where both methods would randomly sample the age structure of the prey species (Klein 1981b).

Klein also suggests that attritional profiles indicate an inability to capture prime age adults, or a predatory preference for easier, (young and old) prey. Hominids scavenging from predators who prefer easier prey might also produce attritional profiles. However, by again invoking the preferential (pre-depositional) destruction of younger individuals by carnivores better able to locate carcasses than hominids, Klein feels that attritional profiles with a high proportion of individuals in the first 10% of life also reflect hunting. Although Klein never states a rubicon for the proportion of very young animals in attritional profiles produced by hominid hunting, he argues that "hunting is probably implied by any archaeological attritional profile in which very young individuals are as well represented as they are in the Klasies giant buffalo profile" (1982a: 157). Very young individuals comprise 63% of the giant buffalo at Klasies (based on Klein 1978). Other larger bovids showing attritional profiles and attributed to hominid hunting by Klein show lower proportions of very young individuals, as low as 36% for blue antelope. According to Klein (1982a), scavenging is a likely explanation for an archaeological attritional profile only if the proportion of very young individuals approaches those seen for the Pliocene paleontological assemblage from Langebaanweg and the Middle Pleistocene paleontological assemblage from Elandsfontein, that is, less than 30%. An important caveat here is that the low proportion of young animals is not taken to be the result of post-depositional destructive forces (e.g., Klein 1982a).

THE SERENGETI CARCASS SAMPLE

Methods

I have employed a four age cohort system for evaluating the context-specific age profiles of the Serengeti carcass sample. I restrict analysis to wildebeest (size 3) to minimize the confounding effects of size and because wildebeest are very abundant in the study sample (see, for example, Lyman 1987b). The age stages are sub-juveniles, juveniles, prime adults and aged adults, as defined by dental eruption and wear (Table 7.2, above), based on visual inspection of mandibles. This was possible for 99 (62%) of the 159 fresh wildebeest carcasses observed in the Serengeti scavenging study.

In order to make my age classification comparable to Klein's published data, it was necessary to estimate that percentage of lifespan elapsed during each stage. I did this by equating my eruption and wear stages to data provided for the blue wildebeest (*Connochaetes taurinus taurinus*) of southern Africa (Attwell 1980) and less complete data for the East African white-bearded wildebeest (*C. t. hecti*, Talbot and Talbot 1963). Table 7.2 shows the age in months estimated by these authors for the various eruption and wear stages used here, and the rough percentage of lifespan these represent, assuming an "ecological" longevity of 18 years (Klein, personal communication, and 1978). Attwell (1980) compared maturation data for the blue and the white-bearded wildebeest, noting that at least up to the beginning of the juvenile stage (as defined here), the latter matures more slowly. Because Talbot and Talbot (1963) do not provide age estimates for the beginning of the prime adult and aged adult life stages, I assume that the delay in dental development and wear relative to the blue wildebeest remains constant throughout life. This exercise permits the ideal catastrophic and attritional age profiles based on 10% increments of lifespan to be transformed into the four stage system employed in the Serengeti study. The transformed profiles in Fig. 7.4 preserve the essential distinctions between catastrophic and attritional mortality: they show reversed proportions of prime age adults and aged adults, and the youngest age class (sub-juveniles representing the first 10% of potential longevity) is similarly abundant in the two profiles (see also Klein 1987:29).

Results

Table 7.3 and Figs. 7.5a and 7.5b show the age distribution of wildebeest carcasses for several contexts in which scavenging can occur in the Serengeti. Following Klein (e.g., 1982a), differences in age distributions between contexts are evaluated using the 2 sample Kolmogorov-Smirnov test, a conservative test in that the differences between age distributions must be extreme for a significant result to emerge. It is appropriate for the ordinal data being examined, and preferable to the *Chi*-square test because of the small sample sizes available for comparison.

Several sets of contexts examined are relevant for evaluating the effect of the "marginality" of scavenging (cf. Binford 1983) on the age distribution of wildebeest carcasses (Fig. 7.5a and Table 7.3). No significant difference is found on the basis of the presence or absence of large mammalian carnivores (lions and/or hyenas) upon my discovery of a carcass, although, as expected, subadults are more likely to be found when lions or hyenas are absent. Likewise, the tissue completeness of carcasses upon discovery does not affect age distributions significantly. Similar results

Table 7.3. Age profiles for wildebeest in the Serengeti carcass sample.

		sub-juvenile	juvenile	prime adults	aged adults	TOTAL
TOTAL	n	14	4	49	32	99
	%	14.1	4.0	49.5	32.3	
lions and/or hyenas:						
present	n	7	2	33	25	67
	%	10.4	3.0	49.3	37.3	
absent	n	7	2	16	7	32
	%	21.9	6.3	50.0	21.9	
completeness:						
with flesh	n	8	1	27	10	46
	%	17.4	2.2	58.7	21.7	
without flesh	n	6	3	22	22	53
	%	11.3	5.7	41.5	41.5	
cause of death:						
natural	n	0	0	6	1	7
	%	0	0	86.7	14.3	
predator	n	6	4	31	26	67
	%	9.0	6.0	46.3	38.8	
habitat:						
riparian woodland	n	1	2	17	4	24
	%	4.2	8.3	70.8	16.7	
plains	n	11	2	30	27	70
	%	15.7	2.9	42.9	38.6	
season:						
dry	n	6	1	21	12	40
	%	15.0	2.5	52.5	30.0	
wet	n	3	2	20	14	39
	%	7.7	5.1	51.3	35.9	
migratory herds:						
absent	n	1	2	10	4	17
	%	5.9	11.8	58.8	23.5	
present	n	8	1	31	22	62
	%	12.9	1.6	50.0	35.5	

Note: Includes only those carcasses (n = 99, or 62%) of the wildebeest sample for which age was determined by dental eruption-wear.

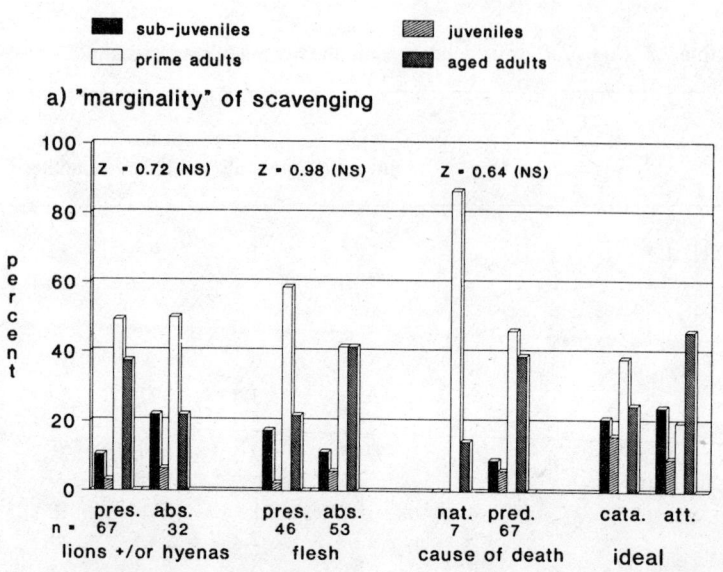

Figure 7.5. Age profiles of wildebeest carcasses encountered in different diurnal contexts in the Serengeti. Part (a) shows three sets of contexts defining the marginality (cf. Binford 1983) of scavenging, including whether lions and/or hyenas were present upon discovery of a carcass, whether the carcass was discovered with flesh or only marrow and head contents remaining, and whether the cause of death was natural or due to predation. Raw data are shown in Table 7.3. The two-sample Kolmogorov-Smirnov Z score is shown for each context. Any $Z<1.96$ is not significant (NS). No significant differences are found for any of the paired contexts. Also shown are the ideal age distributions for catastrophic and attritional mortality taken from Figure 7.4. All but one of the Serengeti carcass sub-samples are more similar to catastrophic mortality in the dominance of prime adults over aged adults, albeit with lower than expected proportions of sub-juveniles and juveniles.

are obtained in a comparison of those carcasses deemed to represent predator kills versus the small number thought to represent "natural" deaths. These results suggest that "confrontational" or "early" scavengers will encounter carcasses with a similar age profile as "passive" or "late" scavengers. Hence, unlike skeletal part profiles, where by definition that

Prey Size and Age Models of Scavenging, Serengeti 141

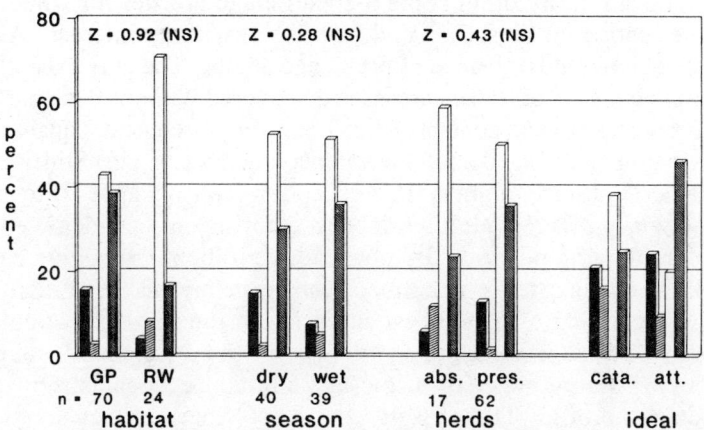

Figure 7.5 *(continued)*. Age profiles of wildebeest carcasses encountered in different diurnal contexts in the Serengeti. Part (b) shows three sets of contexts that define the ecology of scavenging, including habitat, season and whether the carcass was discovered in an area where wildebeest herds were present or absent. Caption under Figure 7.5a discusses statistics.

stage in the consumption sequence at which a carcass is encountered is sensitive to the mode of scavenging (Blumenschine 1986a; cf. Binford 1984, Potts 1983), age profiles remain relatively constant.

Likewise, little variation exists in age structures among different ecological contexts of scavenging (Fig. 7.5b and Table 7.3). I encountered more sub-juveniles and aged adults in plains habitats than in riparian woodlands only because wildebeest calve on the plains, and unusually dry conditions on the plains in the wet season of 1984 may have led to preferential mortality of aged over prime adults. Similar factors probably explain the greater relative abundance of sub-juveniles and aged adults among carcasses found in areas where the wildebeest herds were amassed rather than when they were dispersed (see also Stiner 1990b for other ecosystems). Nonetheless, habitat-related differences in age structure are not significant using the Kolomorgorov-Smirnov test, as is also the case for carcasses encountered during dry and wet periods.

DISCUSSION

The similarity of all context-specific age profiles for wildebeest carcasses shown in Figs. 7.5a and 7.5b is not without interest. All but one profile shows a dominance of prime age adults. The only exceptional subsample consists of those carcasses discovered without flesh but retaining marrow and/or head contents (Fig. 7.5a). In this context, equal numbers of prime adults and aged adults were encountered (see also Stiner 1990b and Chapter 8 of this volume). This exception probably arises from the higher probability of aged adult wildebeest offering only fat-depleted marrow, which in being nutritionally unattractive to bone-crunching carnivores, will increase a carcass' persistence and hence my chance of discovering it.

Scavenging of wildebeest carcasses in the southern woodlands and plains of the Serengeti, therefore, seems to be signaled by age profiles showing a dominance of prime-age adults, i.e., a catastrophic (living-structure) profile. These results contradict Klein's expectation that "...it is difficult to imagine how scavenging could have produced the [catastrophic] eland profile [at Klasies River Mouth], with its abundant prime-age adults, whose carcasses would have been relatively rare (compared to those of very young or very old animals) (Klein 1986:494).

These results raise an interesting paradox. I encountered no major instances of mass mortality in the study. In fact, of the 74 wildebeest carcasses aged by inspection of mandibles and for which cause of death could be determined, only 7 (9.5%) were natural (drowning and old age or disease), *and the remainder predator kills* (Table 7.3).

Here, it is significant that my Serengeti sample did not include carcasses from the dry season refuge of the migratory herds to the west and north of my study area. Wildebeest along with zebra and Thomson's gazelle are the major migratory herbivores in the Serengeti, utilizing the southern plains and plains/woodland ecotone of the central Serengeti around Seronera (my study area) during the relatively wet months of November through May, and the woodlands to the north and west of Seronera during the dry season (e.g., Sinclair and Norton-Griffiths 1979). It is during the dry season that poor forage quality leads to malnutrition, which is more likely to lead to death of aged adults than prime adults. Houston (1979) estimates that over half of all ungulates deaths probably occur during the height of the dry season (August and October), mostly as a result of malnutrition. Hence, wildebeest mortality in the dry season refuge of the Serengeti is probably classically attritional (U-shaped); that is, following the model and associated variation based on the usual effects of disease and malnutrition (e.g., Lyman 1987b; Stiner 1990b) rather than

Prey Size and Age Models of Scavenging, Serengeti 143

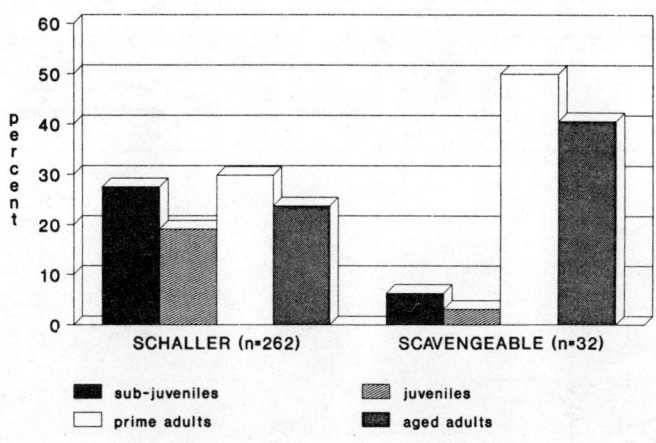

Figure 7.6. Age distribution of wildebeest killed by lions, as recorded by Schaller (1972) primarily through direct observations of predation, and as discovered already dead by me in the daytime. Both profiles are catastrophic in the dominance of prime adults over aged adults, although the carcass sample shows a marked under-representation of sub-adults (*Chi*-square=16.2, df=3, p=0.0011) attributable to their more rapid and complete consumption by lions.

Klein's (1982a) model specifically. However, combined with the fact that Serengeti predators account for only about 30% of annual ungulate mortality in the Serengeti (Houston 1979; see also Schaller 1972, Kruuk 1972), the wildebeest herds that migrate to my study area around Seronera and to the south should be somewhat depleted of aged adults and contain an overabundance of prime adults.

That this is the case is supported by Schaller's (1972) *direct* observations on the age of wildebeest killed by lions, the majority of which were made in the Seronera area to the south of the dry season refuge. Figure 7.6 and Table 7.4 show that Schaller's sample contains a greater abundance of prime adults than aged adults. This catastrophic profile for adult wildebeest is mimicked by my smaller sample of wildebeest killed by lions. In fact the two samples shown in Fig. 7.5 differ only in the relative abundance of subadults to adults (*Chi*-square=16.2, df= 3, p=0.0011). The

Table 7.4a. Age profiles[a] for wildebeest killed by lions in the Serengeti carcass sample and as documented by direct observations of lion predation by Schaller (1972).

		sub-juvenile	juvenile	prime adults	aged adults	TOTAL
Schaller	n	72	50	78	62	262
	%	27.5	19.1	29.8	23.7	
carcass sample	n	2	1	16	13	32
	%	6.3	3.1	50.0	40.6	

[a] See Table 7.3 for the definition of age groups.

Table 7.4b. Correspondence between the age classifications used for the carcass sample and Schaller's kill sample.[b]

age in months at beginning of class	sub-juvenile	juvenile	prime adults	aged adults
Schaller	0	21.6	43.2	108
carcass sample[a]	0	20.5	47	120

[a] See Table 7.3 for the definition of age groups.
[b] Correspondences are as close as Schaller's (1972, Table 45) published age classes (I through X) permit.

under-representation of subadults in my carcass sample is due to the fact that I searched for carcasses only during the day and for individuals already dead rather than by following lions continuously in anticipation of a kill (as Schaller did in most cases). Lions, or scavengers of their kills, had therefore eaten most subadult wildebeest prey before I could discover the carcass, a bias which apparently and importantly did not affect the ratio of prime and aged adults in my sample.

This example shows that in a migratory ecosystem like the Serengeti where lions are abundant, location within the annual range of migratory species and season will affect the age structure of scavengeable carcasses. It also shows that deaths caused by predators, particularly lions, can in certain circumstances produce catastrophic profiles in scavenged assemblages. I have argued that compared to nonviolent deaths caused by malnutrition, predator kills, and particularly felid kills, could have provided a more regular, predictably located and nutritionally higher quality source of carcass foods for a would-be early hominid scavenger in this region of Africa (Blumenschine 1986b, 1987). As such, hominids scavenging in contexts examined in my Serengeti study should encounter carcasses of migratory species in frequencies expected of catastrophic mortality. Populations of residential (non-migratory) species, on the other hand, are more heavily exploited by predators. Carcasses of residential species are therefore more likely to reflect prey age selection habits of predators such as lions, especially since residential prey are relied-upon by residential predators during the season when migratory herds are absent (e.g., Schaller 1972, Kruuk 1972).

Still, it is important to note that the carcass subsamples, while clearly catastrophic in the dominance of prime-age adults, generally also may show a lower combined proportion of young animals than expected by a catastrophic (living-structure) mortality model. However, the proportion of sub-juveniles, which averages to 14.1% in the combined Serengeti sample (minimum=4.2%, maximum=21.9% among the different contexts analyzed), are comparable to those for the Klasies eland (16.5%) and Nelson's Bay Cave bastard hartebeest (24%). These two species show catastrophic profiles, and were interpreted as hunted by Klein. He does so by marshaling selective destruction of teeth of younger animals due to post-deposition compaction and leeching. However, other species from Klasies and Nelson's Bay Cave which are represented by attritional profiles, show a clear dominance of very young individuals over all other 10% increments of life. Actualistic observations from one part of the Serengeti show that one need not invoke drives, traps or repeated natural catastrophes to produce catastrophic profiles in archaeological contexts. Rather, such profiles are available to a diurnal scavenger *at least for wildebeest* as a result of lion predation and other causes of death followed by selective destruction of young individuals by initial consumers.

The apparent absence of any scavenging context that produces wildebeest carcasses in proportions expected by attritional mortality would seem to support Klein's contention that such profiles in archaeological context were not produced by scavenging, but rather by active hunting.

One may argue, however, that dry season natural mortality among migratory species would produce attritional mortality. Klein considers archaeological attritional profiles observed for other taxa (e.g., buffaloes and antelopes) to have been produced by hominid hunting only if the very young age class is abundant. Here, it seems significant that the very high proportion of very young individuals among many species showing attritional profiles at the South African MSA and LSA sites (ranging from 36% to 63% for larger bovids, Klein 1978) is not seen in any of the Serengeti scavenging contexts for wildebeest. Such animals, though preyed upon by lions in frequencies greater than any other age class (28% according to Schaller), simply do not survive for diurnal scavengers (6% in my carcass sample) (see Fig. 7.6).

CONCLUSIONS

The Serengeti data suggest that size and age profiles of scavengeable carcasses are influenced strongly by the death contexts in which scavenging occurs. The available carcass data suggests that size patterns are influenced particularly by the habitat or season-specific structures of predator and prey communities. Size patterns would be further affected by the adaptations of the scavenger of interest, where its realized niche might further restrict the range of variation in carcass sizes procured.

Age profiles may aid in indicating the method(s) hominids used to procure carcasses, but only if the ecological context is also considered and only with regard to the relative abundance of subjuveniles. The Serengeti study confirms and quantifies Klein's and Vrba's argument that sub-juvenile carcasses are less likely to be encountered by a diurnal scavenger than carcasses of adults. If one assumes that hominids practiced a single procurement strategy for acquiring carcasses of a given species, then the Serengeti results support Klein's and Vrba's inference that species showing a dominance of sub-juveniles over other age classes were more likely hunted than scavenged.

However, a dominance of sub-juveniles does not denote that other age classes, particularly adults, were also hunted. Indeed, I have always emphasized that the adults of size 3 animals killed by lions would have provided early hominids with the best scavenging opportunities, simply because younger individuals of these species tend to be rapidly and completely consumed (e.g., Blumenschine 1987).

Criteria other than the relative abundance of sub-juveniles are required to determine whether adult individuals were hunted or scavenged.

Although most contexts of wildebeest carcass availability examined in the Serengeti produce catastrophic, prime-adult dominated profiles, I have argued that other contexts of wildebeest mortality in the Serengeti that were not observed might produce attritional (U-shaped) profiles. For the majority of archaeological assemblages, detailed reconstructions of the site's ecological context are required if age profiles are to be relied upon for determining whether hunting and/or scavenging took place. Because such reconstructions are fraught with their own difficulties, it is clear that other zooarchaeological criteria, specifically skeletal part profiles and surface modifications on bones, must be integrated with mortality profile data if the carcass procurement strategies practiced by hominids are to be ascertained.

ACKNOWLEDGMENTS

Funding for the field work was provided by the National Science Foundation, the L. S. B. Leakey Foundation, and the Regents of the University of California. I am grateful to the Tanzania Commission for Science and Technology, the Serengeti Wildlife Research Institute and the Tanzania Antiquities Unit for permission to conduct the field research and for cooperation throughout. Charles Saanane provided valuable field assistance.

NOTE

1. It is interesting to note that the under-representation of size 1 and 2 animals in the carcass sample would be even more extreme if the size distribution of Serengeti herbivore deaths rather than live animals were used. This is due to the fact that populations of smaller species have a relatively high turnover rate. For example, Houston (1979) reports that Thomson's gazelle (size 1) have a 20% annual adult mortality compared to 12% for wildebeest, 10% for zebra (both size 3) and 4% for elephant (size 6).

8

An Interspecific Perspective on the Emergence of the Modern Human Predatory Niche

Mary C. Stiner

INTRODUCTION

Several aspects of the European Middle Paleolithic archaeological record suggest that Neandertal lifeways differed significantly from those of later, anatomically-modern humans (e.g., Binford 1983; Gamble 1986; Mellars 1989). However, little can be said about how these differences might reflect adaptive change, because so few independent measures currently are available for assessing ecological "distance". With this issue in mind, I have chosen to investigate possible shifts in the predatory niche of *Homo sapiens* from an interspecific point of view. The study is conducted from the perspective of niche theory and examines hominids' ecological relationships with both their prey and coexisting predators.

Niche theory addresses the processes by which animal communities are assembled and, with time, rearranged (e.g., Diamond 1975; MacArthur 1968; MacArthur and Levins 1967). The foraging niches of species are defined by the resources that animals depend upon and by how and when they use them. Because the niche boundaries of species with similar foraging interests are thought to be contiguous, knowing about one species can help to understand another. This approach is especially well-suited to comparisons of the subsistence systems of Neandertals and Upper Paleolithic peoples because both were ungulate predators, both sometimes used caves for shelter, and both coexisted with a variety of large carnivores. These relationships present an arena in which evolutionary changes in human subsistence adaptations and land use can be examined. Moreover, because the Carnivora have undergone relatively little evolutionary change in the Upper Pleistocene, their resource use habits can serve as more or less independent constant for measuring change in *Homo sapiens*, a predator whose adaptations were less stable across the same time range.

Here I discuss only two dimensions of foraging systems. The first dimension, based on mortality patterns in ungulate death assemblages, represents prey age selection habits of hominids and nonhuman predators. The second, based on anatomical representation in shelter faunas, represents food transport habits of bone-collecting predators (hominids, hyaenas and wolves). Both variables are fundamental to ecological models of foragers, as they reflect how predators search for and procure prey, their decisions about what food items to move and where to move them, as well as certain constraints on processing (Stiner 1990b, 1991b). Of special interest here is the relative emphasis on hunting versus scavenging at the assemblage level.

THE STUDY SAMPLE

The comparisons emphasize large predators, primarily those that deliberately carry bones into shelters (dens, shallow caves, rockshelters). Artiodactyl prey (deer, cattle and bison in temperate settings; antelopes and buffalo in arid tropical settings) comprise most of the prey sample for the mortality analyses, because they are common in the diets of large terrestrial predators in most environments. Analysis of bone transport is restricted primarily to those ungulates in the intermediate body size range in order to minimize the influence of carcass "persistence" at procurement sites on the transport options of predators.

The analyses, by design, cross taxonomic boundaries in both predators and their prey. It meanwhile must be acknowledged that some researchers have observed considerable variation in food transport decisions by humans according to the species procured (O'Connell *et al.* 1988a), and this undoubtedly is relevant to understanding finer-scale variation in human foraging practices. It remains true, however, that one cannot hope to isolate general ecological principles concerning niche in different time periods and environments without also relaxing taxonomic restrictions to some degree. This does not pose a contradiction to other research; it is simply a different scale of observation, in this case, chosen to deal with a community level problem.

Each "case" represents one (or two similar) prey species from an archaeological assemblage, or from a specific time frame and study area in modern situations. Holocene archaeofaunas from North America and a wide variety of modern and Pleistocene bone assemblages generated by nonhuman predators in the New and Old Worlds serve as control cases for the study. Most of the Paleolithic archaeological cases are from a series of caves and rockshelters in Latium, a province of west-central Italy (Table 8.1, see also Stiner 1990a). The Mousterian (Middle Paleolithic) material comes from 4 stratified cave sites dating between 110,000 to 35,000 B.P.: Grotta Guattari, Grotta Breuil, Grotta di Sant'Agostino and Grotta dei Moscerini. The Upper Paleolithic sample includes cases from Grotta Palidoro, Riparo Salvini and Grotta Polesini in the same general region of Italy and from La Riera cave in northern Spain, an area with a similar climate (20,500-8,000 B.P., Altuna 1986, Straus 1986; see Stiner 1991b for data conversions).

Taphonomic analyses were essential for establishing the connections between faunal assemblages and bone-collectors in the Upper Pleistocene cases. These findings, along with descriptions of the basic data and background information for the sites, are presented in other publications (see Stiner 1990a, 1990b, 1991a, 1991b). With the exception of the Salvini material, the state of preservation for the Italian archaeological faunas is generally very good, due to the limestone formations in which the faunas occur. However, this in itself cannot taken as a guarantee that no *in-situ* bone destruction has taken place. The worst potential biases in anatomical representation due to bone decomposition are circumvented by considering only the more durable parts of prey skeletons. The possibility of selective deletion (e.g., of epiphyses) by predators feeding *at shelters* or recovery biases by earlier excavators were controlled for in several ways, the descriptions of which are also covered in the references cited above.

METHODS

Two very different kinds of faunal data are used in the comparisons. Mortality patterns, which serve as indicators of prey age selection, are based on *tooth* eruption-wear data. A simple measure of anatomical representation — the proportion of head and horn parts to limb bones above the foot — is used as an indicator of food transport choices. The anatomical variable is based exclusively on *bone* MNE counts; teeth were not used

Table 8.1. Background information for the Italian Paleolithic sites.

shelter site	cultural association	excavator/ research group	background references
Grotta Polesini	late UP (10,000 B.P.)	Radmilli	Radmilli 1974
Grotta Palidoro	UP (15,000 B.P.)	IIPU	Bietti 1976-77a,b,c Cassoli 1976-77 Segre 1976-77
Grotta di Sant'Agostino	late MP (45-55,000 B.P.)*	Tongiorgi	Laj Pannocchia 1950 Tozzi 1970 Stiner 1990a,b, 1991b
Grotta Breuil	late MP (35-50,000? B.P.)*	A. Bietti (IIPU, U. Roma)	Bietti et al. 1988 Taschini 1970 Bietti & Stiner n.d. Stiner 1990a,b, 1991b
Grotta dei Moscerini	MP (110-70,000 B.P.)*	IIPU	Vitagliano 1984 Stiner 1990a,b, 1991b
Grotta Guattari	MP (78-51,000 B.P.)*	IIPU	Blanc & Segre 1953 Piperno 1976-77 Taschini 1979 Stiner 1990b, 1991a, b

UP - Upper Paleolithic, MP - Middle Paleolithic
IIPU - The Istituto Italiano di Paleontologia Umana research group 1939-1955

* Dates determined by ESR method on large ungulate teeth and by U-Series method on horizontal stalagmite layers by H. P. Schwarcz and R. Grün (Schwarcz et al. n.d.).

for determining cranial MNE, because they are more resistant to most destruction processes than bone. The analytical methods specific to each class of data are described below.

Mortality Patterns Based on Tooth Eruption and Wear

The mortality patterns of cervids and bovids in the archaeological cases are constructed from deciduous and permanent occlusal tooth eruption and wear sequences of the lower 4th premolar unless otherwise specified. The wear stage criteria correspond to collapsed versions (Fig. 8.1) of systems developed by Payne (1973), Grant (1982) and Lowe (1967). I favor the P4 sequence for several reasons: (1) the tooth is easily distinguished from the other cheekteeth; (2) it is located at or very near the fulcrum of wear (as is the M1, see Gifford-Gonzalez, this volume) and will be completely worn away with advanced age; and, (3) it effectively represents the complete lifetime of individuals if data for deciduous and permanent counterparts are combined.

The age frequency data are divided into three age cohorts: juvenile, prime adult and old adult. These age stages do not represent equal fractions of the maximum potential lifetime, and instead are based on major life history transitions typical of most ungulates. No attempt is made to estimate the real age of individuals except at the deciduous/permanent tooth replacement boundary. The three age stages correspond to differing susceptibilities to attritional death, as well as to significant differences in body composition. Particularly important from a predator's perspective is the fact that adult cohorts offer substantially higher fat returns on the average than do juveniles in wild populations (for discussions of human responses to these differences in prey, see Binford 1978; Speth and Spielmann 1983; Yellen 1977; and Speth, this volume). Because only three age stages are considered, the age-frequency data can be converted to percentages and plotted as single coordinates on a triangular graph.

The transition between juvenile and prime adult age groups is very clear, since it corresponds to the time when the milk tooth is replaced by a new, permanent one. Because the permanent tooth supplants the deciduous counterpart by pushing it up and out of the jaw, a permanent tooth must show some evidence of occlusal wear, however slight, in order to be counted. The prime-old age boundary is based on how much of the tooth is left relative to the complete form and the appearance of the

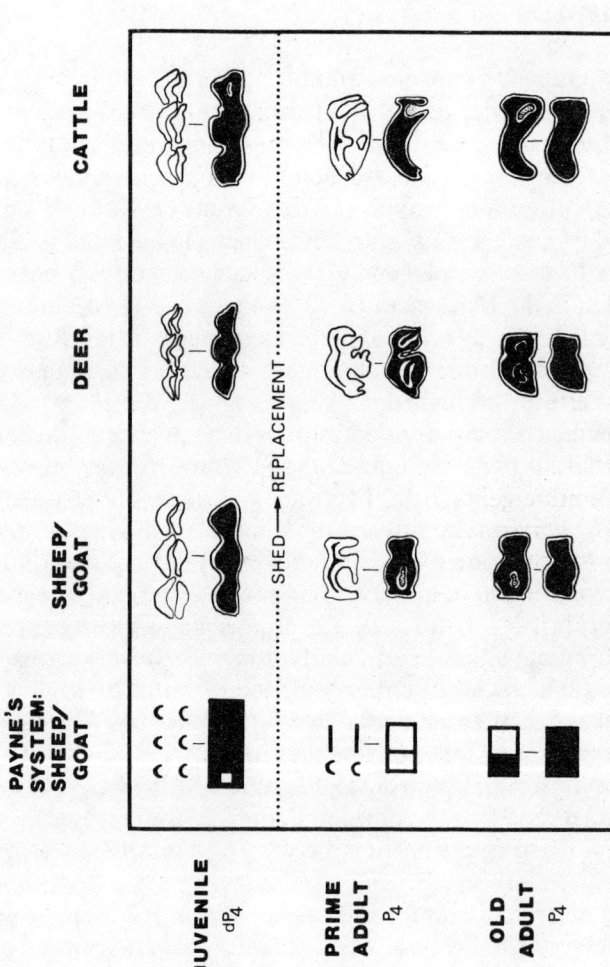

Figure 8.1. Occlusal wear stages according to the three age system for deciduous and permanent lower 4th premolars of sheep/goat, cattle and deer. Drawings of domestic sheep/goat and cattle teeth adapted from Grant (1982), deer teeth based on Lowe (1967) and Stiner's comparative drawings for red, fallow, white-tailed, and mule deer.

occlusal surface. The old age category begins when more than half of the tooth crown is worn away, and it theoretically ends when animals run out of tooth (Fig. 8.1). That animals can actually reach such an advanced state of tooth wear is confirmed by a few individuals in several of the archaeological assemblages used for this study.

Transposing data from other archaeological and modern wildlife studies, which employ diverse methods and styles of presentation, is less straightforward. In wildlife studies, aging is based on observed reproductive state and/or tooth wear to the extent that it is visible above the gum line. Some ambiguity in converting the age data cannot be avoided when using other published sources. However, the system employed here minimizes the potential impact of these errors because only three categories are used, and the boundaries between age groups are based on life history transitions in prey that can also be observed from external characteristics.

The physiological correlates of the three age categories can be summarized in the following way. In the juvenile age group, which is defined by the presence of deciduous dentition and/or emerging but unworn permanent teeth, the zones of hemopoiesis (red marrow) are not yet restricted to the cancellous tissues of the axial skeleton and proximal upper limbs (e.g., Crouch 1972 on humans; but this developmental process applies to mammals more generally, B. Snyder D.V.M., Rio Grande Zoo). Little to no fat (yellow marrow) is stored in the medullary cavities of the long bones. The social and neurological immaturity of juveniles make them more vulnerable to harsh environmental conditions, disease and predators. These facts, along with the naturally great abundances of juveniles in live populations, are responsible for their great abundance in both living-structure and U-shaped ("attritional") mortality patterns.

The prime adult age group consists of sexually mature individuals in which the permanent tooth is in place and in wear. This stage represents the peak reproductive years in ungulates, although the beginning and end of this phase vary according to gender. It is important to note that the juvenile/prime adult transition is based on the physiological capability to reproduce and *not* on realized productivity. The former is a better reflection of changes in the ability to store fat. Separation of the juvenile and prime age groups in this study is based on females' entry into reproductive phase. In temperate and colder environments, healthy prime-aged adults develop substantial fat stores in the limb bones in advance of the cold season and, in the case of dominant males, in preparation for the fall rut.

The old age stage begins at roughly 61-65% of the maximum potential lifetime, although variation in longevity among populations introduces some confusion if this criterion is used alone. Tooth wear information was given priority whenever possible. An average (population level) decrease in productivity is associated with this age group, although this does not affect every individual equally (e.g., Berger 1986 on feral horses; Clutton-Brock et al. 1982 on red deer; Geist 1971:284 on wild sheep and reindeer; Leslie and Douglas 1979:23 on desert bighorn; Sinclair 1977: 226-228 on African buffalo).

Anatomical Representation Based on Bone MNE Counts

Anatomical representation of shelter faunas is based on bone MNE counts. The rationale for using the bone MNE data and how the determinations were made are presented in other publications (e.g., Stiner 1991b). The MNE counts exclude carpals, tarsals other than the calcaneus and astragalus, and all teeth, because the uniformly dense structure of these elements can inflate their relative abundances in archaeological contexts, and because they arrive in shelters as non-nutritious riders attached to larger food-bearing bone elements.

The MNE data are examined in terms of two simple ratios. The first ratio, called tMNE/MNI, serves as an index of anatomical completeness of prey remains. It is obtained by dividing the *total* number of skeletal elements (tMNE) for a species in an assemblage by the minimum number of individual animals (MNI) represented. The ratio represents the average quantity of substantial bony parts transported to shelters per carcass source. The second ratio, called H+H/L, is the sum of horn/antler and head MNE counts for a species in an assemblage divided by the MNE for all major limb bones above the phalanges. The H+H/L ratio accounts for roughly 50% of the variation observed in the faunas collected by all of the predators considered. While biasing effects of bone attrition on anatomical representation cannot be wholly excluded in these comparisons, I have shown elsewhere that these processes do not explain much of the variation observed among the faunas (e.g., Stiner 1991a, 1991b). Regressions of H+H/L against tMNE/MNI trace the relationship between hunting and scavenging strategies as expressed by the anatomical parts most commonly represented in shelters.

MORTALITY PATTERNS GENERATED BY NONHUMAN PREDATORS

A predator's feeding "choices" are conditioned by what living prey populations have to offer and, in some cases, by age-specific susceptibility of individual prey to that predator's procurement techniques (see also Chapter 1). For these reasons, mortality patterns generated by most predators mimic either those arising from attritional factors such as disease, malnutrition and accidents (U-shaped patterns, Fig. 8.2b), or the patterns represent a more or less "random" sample of the living population from which they were derived (Fig. 8.2a). Prime-dominated mortality represents a third major class of mortality pattern relevant to the study of humans (Fig. 8.2c). While not described in the ecological literature, prime-dominant mortality patterns are documented for some modern human hunters (e.g., Nunamiut Eskimo, Stiner 1990b, see also Binford 1978) and in some archaeological records (e.g., Frison 1984; Levine 1983; Stiner 1990b; Todd 1983; Todd and Hofman 1987).

Because mortality patterns most often are considered in a two-dimensional format, some explanation of their correspondences in the three-dimensional graphs used in this presentation is needed. Figure 8.3a defines the three age axes and relates areas of the graph to general classes of mortality pattern. Each of the three axes *bisects* the triangle and ranges from 0 to 100%. Figure 8.3b shows the relative placement of the major classes of mortality patterns, along with the expected range of variation associated with each.

Variation associated with any type of mortality pattern is seldom dealt with effectively in archaeological studies (but see Cribb 1987; Lyman 1987b). Comparisons tend to rely exclusively on idealized models when, in fact, such models at best approximate the mean for a 'family' of profile patterns. Knowing the typical range of variation for each model is critical for assessing "difference" among real cases, especially when working with static models. The range of variation is also likely to be of ecological or evolutionary significance in the study of any predator. For both of these reasons, I have constructed expected variation ranges based on empirical observations with a generous margin subsequently added (Stiner 1990b: 319-323).

The two most common mortality patterns (U-shaped and living-structure) occupy the lower central areas of Fig. 8.3b. The corners of the graph correspond to other classes of mortality, each strongly biased toward the age group designated. Old individuals are naturally rare in living populations, so that cases occurring anywhere in the upper half of

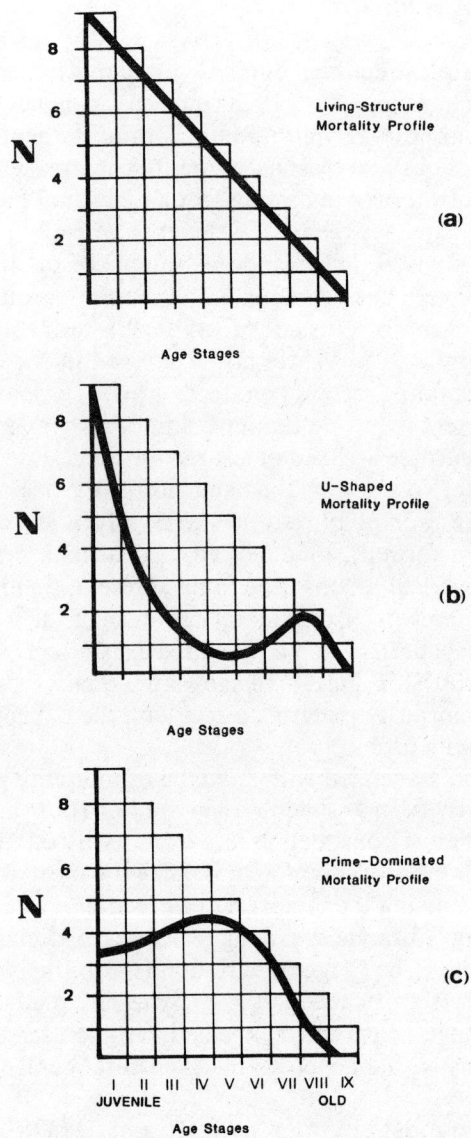

Figure 8.2. Three idealized mortality profiles found in nature, superimposed on a stable living population model (from Stiner 1990b).

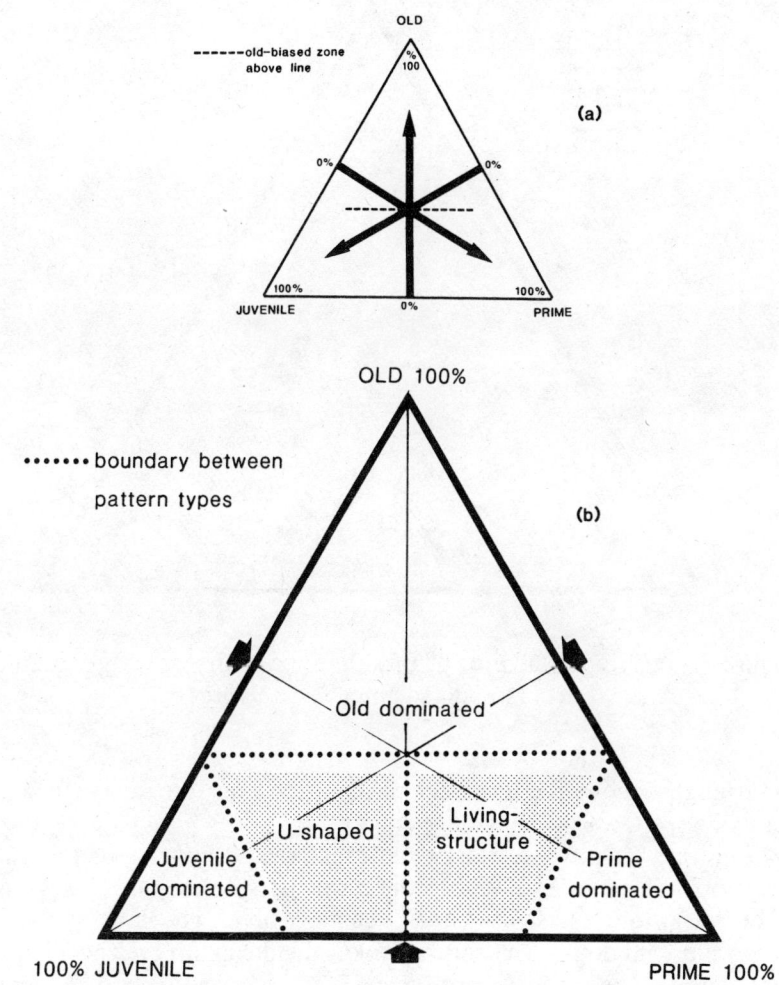

Figure 8.3. Triangular diagrams (a) defining the three age axes, and (b) relating areas of the graph to general classes of mortality pattern.

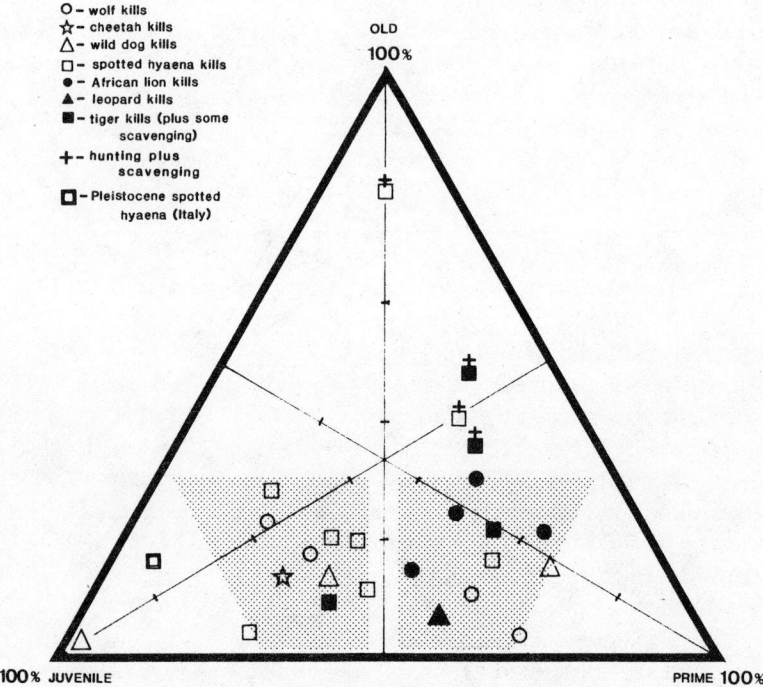

Figure 8.4. Triangular plot of ungulate mortality cases generated by modern cursorial and ambush carnivores.

the triangle (above the level where the three axes intersect) are significantly biased towards the old age cohort.

Figure 8.4 shows mortality patterns created by various nonhuman predators. The open symbols denote carnivore species generally classified as cursorial/long-chase predators, such as wolves, cheetahs, spotted hyaenas and wild dogs. The solid symbols denote carnivore species generally classified as ambush/short-chase predators, which includes most of the cats (data are summarized in Stiner 1990b).

Most cursorial predators — species that run down their prey — fall in the lower left half of the graph. They *tend to* kill young or weak individuals, usually producing U-shaped mortality patterns in prey death assemblages. The classic ambush or stalking predators are represented in the lower right half of the graph. For them, prey age selection is determined by encounter, and the cumulative mortality patterns usually mimic the age structure of live populations. The distribution of cases illustrates well-known, although general, behavioral and niche differences between cursorial and ambush carnivores (e.g., Bertram 1979:228; Ewer 1973; Kruuk 1972:274-283; Schaller 1972:392).

There are two interesting exceptions to these generalizations. First, cursorial species sometimes behave like ambush predators by using natural corrals, lake ice, thick vegetation and other landscape features to disadvantage prey. These conditions are documented for 3 of the 4 aberrant "cursorial" predators cases occurring in the lower right half of Fig. 8.4;[1] the circumstances of the fourth case are not known. Such behavior may not be particularly common in any cursorial species, but it is occasionally documented in nearly all of them (e.g., Asiatic dhole [Fox 1984]; wolf [Bibikov 1982, Miller 1975]; spotted hyaena [Kruuk 1972]).[2] Cursorial species behave like ambush hunters *only* with the aid of natural environmental features, and, in these circumstances, the cumulative result is a living-structure mortality pattern. This shows that prey age "choice" is not intentional on the part of individual predators, but rather is an artifact of the way in which a strategy allows the predator to get close enough to prey animals to grab and kill them.

The second set of exceptions in Fig. 8.4 is represented by two tiger and two spotted hyaena cases, all of which display old-age biased mortality patterns and are distributed above the level where the three age axes intersect. Each example involved a relatively high dependence on scavenging, as indicated by Schaller's (1967) discussion of his sample and field methods in the Indian tiger study, and by Tilson et al. (1980:49) in their study of Namib spotted hyaenas. These predators existed at or near the top of the local carnivore hierarchy, and they scavenged ungulates perishing primarily from "non-violent" causes (including eventual infection from predator-caused injury). The patterns associated with both cursorial (spotted hyaenas) and ambush predators (tigers) converge on the same area of the graph as the importance of scavenged food sources increases.

The fact that attritional death is the most common mortality pattern in nature may explain the relationship between the old bias and scavenging behavior. Recall that prime-age animals are much less likely than young and old animals to succumb to attritional death. Many more juveniles than old adults may die simply because young animals are more common in the live population. However, more thorough destruction of juvenile carcasses by primary feeders (e.g., Schaller 1967, 1972; Haynes 1980, 1982; Blumenschine 1986b, 1987), the difficulty of locating small carcasses, and faster rates of biochemical deterioration due to low volume relative to surface area (e.g., Lyman 1984) often prevent scavengers from benefiting from juvenile deaths. As a result, old adult carcasses, which are more persistent due to larger size and harder tissues, often become available to scavengers even though not as many die within a given period. Tilson *et al.* (1980:47,49) make explicit links between scavenging and the old-age bias in their study of spotted hyaenas in the Namib Desert, and Schaller (1967) implies a similar explanation in his study of Indian tigers.

Observations relating procurement strategies to spatial properties of food supply point to significant differences in levels of mobility and search time required for hunting as opposed to scavenging (see also Houston 1979). Links between highly dispersed food supplies and old-biased prey selection are documented for both spotted hyaenas (Tilson *et al.* 1980; Mills 1984b) and lions (Mills 1984b) in arid southern African environments. This suggests that environmental stress and degree of food dispersion, whether acute or chronic, can exert a strong influence on foraging responses of species that both hunt and scavenge. The documented association of the old-age bias and scavenging in cursorial and ambush nonhuman predators suggests a more general mortality model that could be helpful for identifying scavenging by hominids, especially if carcasses were obtained from non-violent sources of death and/or cursorial predator kills (but see also Blumenschine, this volume, for scavenging opportunities arising in habitats where large cats are abundant).

PREY AGE SELECTION BY HOLOCENE AND PALEOLITHIC HUMANS

Figure 8.5 illustrates the distribution of white-tailed deer, caribou and bison mortality cases from Holocene human sites in North America (circa 12,000 B.P. to present, data and sources presented in Stiner 1990b). Figure 8.6 shows the distributions of red deer and aurochs (wild cattle) cases from Mousterian and Upper Paleolithic shelter sites in west-central Italy. Only the dominant species in the archaeological assemblages are considered.

The Holocene cases in Fig. 8.5 cluster in the lower right half of the graph, forming a continuum between *nonselective* living-structure mortality and *selective* prime-dominated mortality. The plotted cases represent Paleoindian, early Archaic and Protohistoric bison hunters of the North American Plains (Frison 1984; Speth 1983), historic Nunamiut Eskimo reindeer hunters (Binford's collections, and 1978), Middle Mississippian deer hunters (B. Smith 1975), and modern trophy hunters using

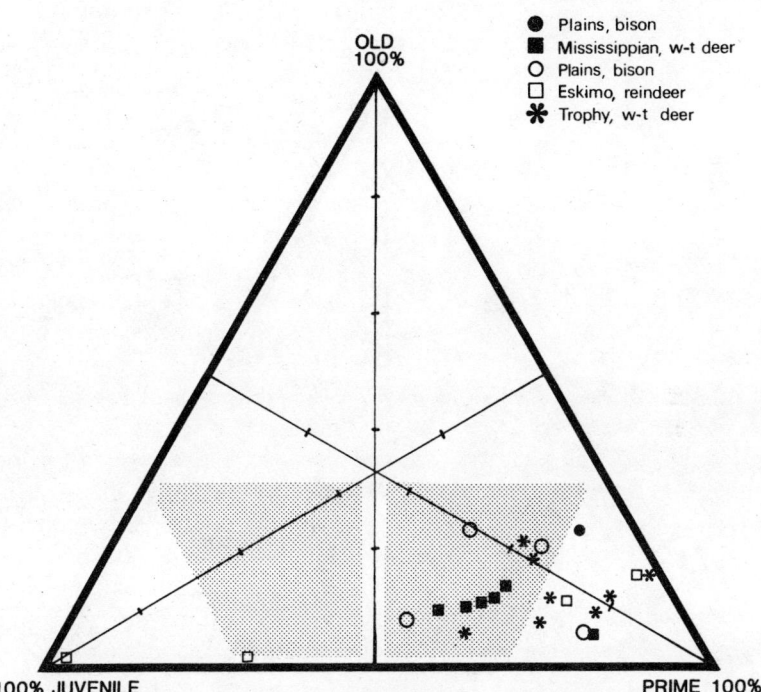

Figure 8.5. Triangular plot of mortality patterns for cervid and bovid remains in North American Holocene archaeofaunas and patterns generated by modern trophy hunters.

firearms (Woolf and Harder 1979). The two aberrant Nunamiut Eskimo cases occurring in the lower left corner of the graph involved deliberate (also selective) predation on caribou calves to obtain soft hides for clothing (Binford 1978, and unpublished field notes). The distribution of the Italian Upper Paleolithic cases in Fig. 8.6 (open circles) is entirely consistent with the Holocene pattern. In contrast, the Middle Paleolithic cases (open triangles) occur over a much larger area of the "ambush half" (right side) of the graph.

Two general observations based on these data merit elaboration. One is the relatively greater variation in prey age selection apparent in the Middle Paleolithic. The second concerns the behavioral and ecological

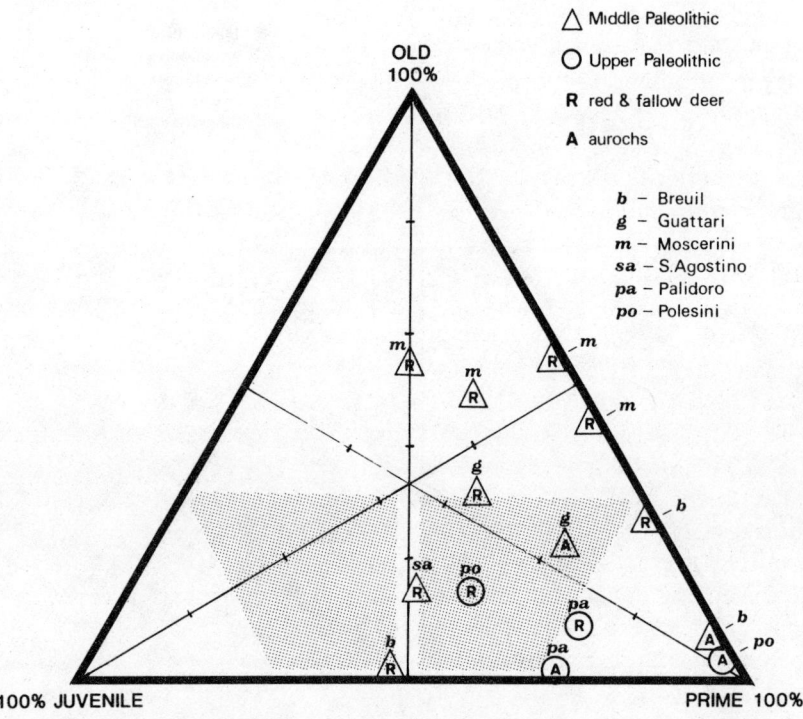

Figure 8.6. Triangular plot of mortality patterns in cervid and bovid remains in Paleolithic Italian shelters.

implications of mortality patterns characterizing modern human hunting practices.

Variation among the Mousterian cases is far in excess of that for all Upper Paleolithic and Holocene cases combined. This variation segregates according to site; essentially the same procurement practices are repeated throughout any given stratigraphic sequence. Some Mousterian cases fall within the modern human scatter (from Grotta Breuil and Grotta di Sant'Agostino), but the old-biased mortality cases from two other Mousterian sites (Grotta dei Moscerini and Grotta Guattari) contribute to a much larger spread of points. The old-biased cases are distributed in the upper right part of the graph, occurring in the same area of the graph as the nonhuman predator cases associated with high levels of scavenging. These data suggest that at least two distinct ungulate procurement strategies, hunting and scavenging, were very important in the Mousterian of this small region of Italy. Only hunting dominates in the Upper Paleolithic and Holocene cases.

As regards hunting specifically, humans appear to be the only agencies that *regularly* produce prime-biased mortality patterns in prey death assemblages, making this kind of pattern more specific to cause than the other mortality patterns described previously (see also Stiner 1990b). Ethnographic and historic accounts (e.g., Benedict 1975; Egan 1917; Regan 1934; Steward 1938; Tatum 1980), ethnoarchaeological studies (Binford 1978) and some archaeological evidence (e.g., Frison 1984; Todd 1983; Todd and Hofman 1987) generally associate prime-dominant harvesting with "selective" ambush hunting. In modern contexts, the prime-dominated mortality pattern occurs when trophy hunters use firearms (e.g., Sowls 1984; Woolf and Harder 1979).

The data presented indicate no basis for suggesting hominids were cursorial *hunters* in any capacity (cf. Klein 1978, 1981a, 1981b, 1982 on South African MSA faunas, but also see 1987, 1989 for revised view), although it is at least conceivable that they could produce U-shaped mortality patterns in sites by scavenging. However, this would require that scavenging opportunities were (1) from cursorial predator kills and/or nonviolent deaths and (2) the carcasses of juveniles were sufficiently large to persist on a landscape (see, for example, discussion by Blumenschine, this volume).

PREY AGE SELECTION AND PREDATORY NICHE

To simplify the comparisons so that the habitual behaviors of predators may emerge more clearly, only median values for groups of "like" cases are considered in Figs. 8.7 and 8.8 (and Table 8.2.) Here, predators are classified according to how they most often exploit the age structure of prey populations. Three clusters are apparent in Fig. 8.7, and they are roughly equidistant from one another. Variation associated with each set is not plotted; suffice it to say there is some overlap between the ranges associated with each cluster, as shown in previous graphs.

The first cluster, on the lower left portion of Fig. 8.7, consists of medians for cursorial carnivores, including wolves (Ncases=3), spotted hyaenas (Ncases=8) and African wild dogs (Ncases=3). The second cluster, high on the right, contains medians for two classic ambush predators, lions (Ncases=4) and tigers (Ncases=4). Leopards are not considered because only one case was available. The third cluster, on the lower right,

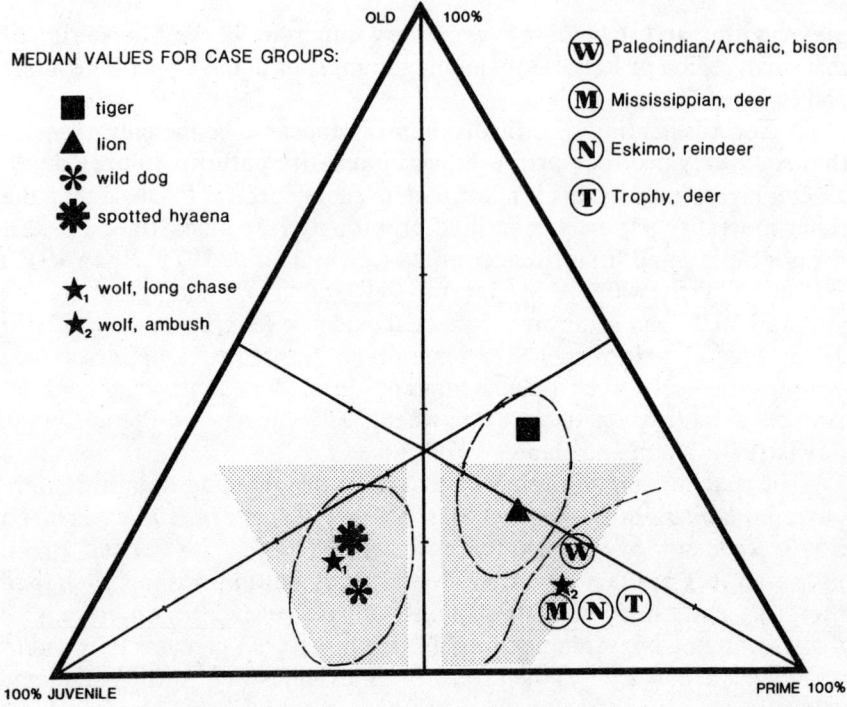

Figure 8.7. Triangular plot of median mortality values for modern predators grouped by strategy class or (in human cases) environment and culture.

consists of medians for Holocene human hunters and, rarely, cases in which cursorial predators used landscape features to entrap prey (wolf, Ncases=2, but see footnote 1). As is true for the big cat group, this third cluster also represents ambush predatory behaviors, but of a different, more biased character relative to natural processes of recruitment and attrition in mammalian populations.

The Holocene human cases are grouped by culture, time period and geographical origin. The median values represent Paleoindian, Archaic and Protohistoric bison hunters of the North American Plains (Ncases=4), Nunamiut Eskimo reindeer hunters (Ncases=5), Middle Mississippian deer hunters (Ncases=6), and modern trophy hunters using firearms (Ncases=8). The medians are remarkably similar to one another in spite of the diverse environments and technologies represented, and all show some bias towards prime adult prey.

Emergence of Modern Human Predatory Niche 167

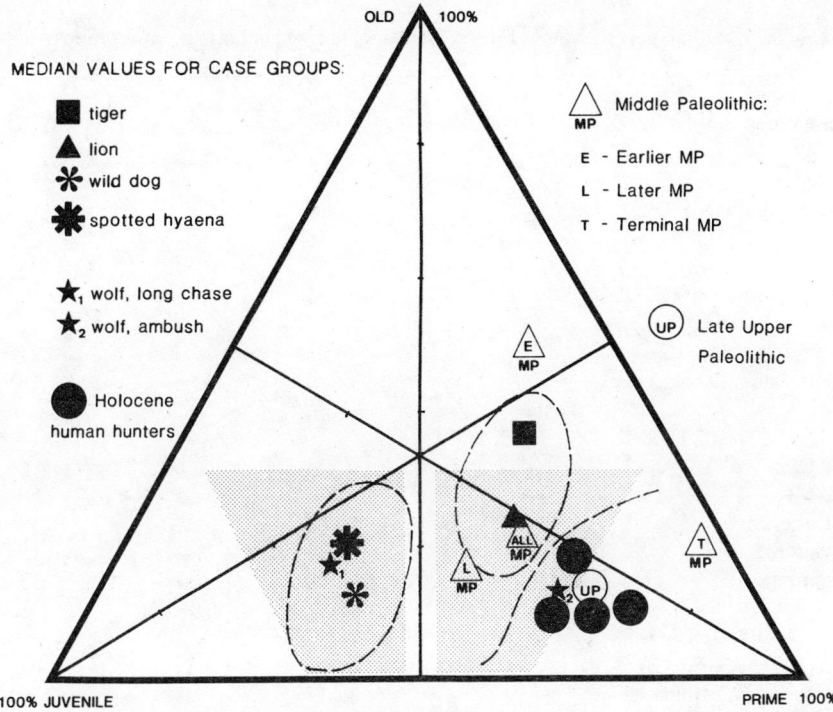

Figure 8.8. Triangular plot of median mortality values for various modern and Paleolithic predator groupings.

Figure 8.8 repeats the same graph, but with the addition of the Italian Paleolithic median values. The dark symbols denote modern predators, including Holocene humans, whereas the open symbols denote Mousterian (MP) and late Upper Paleolithic (UP) medians. The median for the late Upper Paleolithic (Ncases=10) is nested in the center of the Holocene human cluster. The combined Mousterian median falls closest to the big cat cluster, just below the African lion median. When sorted by chronological age, the three Mousterian sub-medians are widely scattered on the graph, and only the youngest case group (circa 40-35,000 B.P.) shows any indication of being dominated by prime-aged prey.

I subdivided the Mousterian case groups based on distinct patterns of anatomical composition and approximate chronological age (see Stiner 1990a:Appendix 1; Schwarcz et al. n.d.). The earlier cases in the Middle Paleolithic sample (E-MP, Ncases=6) date to roughly 110-65,000 years ago, are consistently biased towards old prey (Grotta dei Moscerini level

Table 8.2. Median percent values for the three ungulate age cohorts by case groups.

case group	N cases	JUV	PRIME	OLD
NON-HUMAN PREDATORS				
Indian tiger (A)	4	.15	.41	.29
African lion (A)	4	.24	.47	.23
African wild dog (C)	3	.52	.34	.13
spotted hyaena (C)	8	.47	.29	.21
wolf, long chase (C)	3	.53	.30	.22
wolf, "ambush" behavior[a]	2	.30	.63	.15
PALEOLITHIC HOMINIDS, ITALY				
all UP	10	.23	.66	.15
all MP	19	.26	.54	.20
earlier MP	6	.12	.40	.47
later MP	8	.36	.47	.15
terminal MP	5	.00	.71	.13
HOLOCENE HUMANS, NORTH AMERICA				
modern trophy hunters	8	.17	.70	.11
historic Nunamiut hunters	5	.26	.68	.05
(without "skin" hunting)	3 [b]	.16	.72	.12
Mississippian hunters	6	.28	.61	.09
Paleoindian/Archaic hunters	4	.20	.58	.14

(A) - generally classified as an ambush predator
(C) - generally classified as a cursorial predator

[a] Two distinct prey age selection behaviors are documented for North American and Eurasian wolves. This species usually produces U-shaped mortality patterns in prey through the cumulative effects of hunting. However, some environments offer vegetational or geomorphological features that allow packs to ambush or "corral" prey, as for the two cases listed here.

[b] Situations in which Nunamiut hunters specifically sought juvenile reindeer skins for clothing are omitted.

Note: The *median* values for percentages in the three age cohorts do not always equal 100, because the medians are for multiple cases and the value for each age stage is calculated independently.

groups M2, M3, M4 & M6 and Grotta Guattari G4-5) and, as will be shown in the next section, consist almost exclusively skulls and mandibles of red deer and aurochs (see also Stiner 1991a, 1991b). The E-MP median fits the expectations for scavenging, based on analogous patterns for nonhuman predators.

The second median is for later Middle Paleolithic cases from levels S0-S4 of Grotta di Sant'Agostino (L-MP, Ncases=8), dating to approximately 65-40,000 years ago. This group suggests nonselective ambush hunting and associates with anatomical patterns rich in all skeletal elements except the spinal column and ribs. Here, hominids clearly had full access to all meaty parts of carcasses procured.

The terminal Middle Paleolithic median consists of cases from Grotta Breuil (T-MP, Ncases=5), dating to 40-35,000 years ago. Anatomical representation in these cases is similar to that of the L-MP group from Sant'Agostino, but the age structures are quite different. The T-MP pattern exhibits a strong bias favoring prime adult prey. The mortality patterns represented by T-MP are as extreme in this respect as any generated by Upper Paleolithic humans in Italy (L-UP) and various Holocene human cultures in North America.

FOOD TRANSPORT AND PROCUREMENT STRATEGIES

Anatomical representation in prey remains carried to shelters by predators will not necessarily diagnose particular bone-collecting species. Instead, it reflects the predominant foraging strategy employed, as well as seasonally-conditioned responses to the quality and abundance of food. Only some predators *deliberately* collect bones in shelters — a criterion that includes humans, canids and hyaenids, but generally excludes the cats (even leopards). As with the mortality analyses, I begin with interspecific comparisons of modern predators, drawing relationships between foraging strategies and transported prey body parts. I then proceed to comparisons of Paleolithic cases. Note that the observations made here primarily concern medium ungulates and may or may not apply to the treatment of significantly larger or smaller prey taxa.

Relationships between foraging strategies and food transport habits can be examined in terms of the average amount of food transported from carcass sources (anatomical completeness, tMNE/MNI) and anatomical content (H+H/L). Dividing tMNE by MNI standardizes the quantity of skeletal parts in shelters relative to the average number of foraging "events" (carcasses). As explained previously, anatomical content is

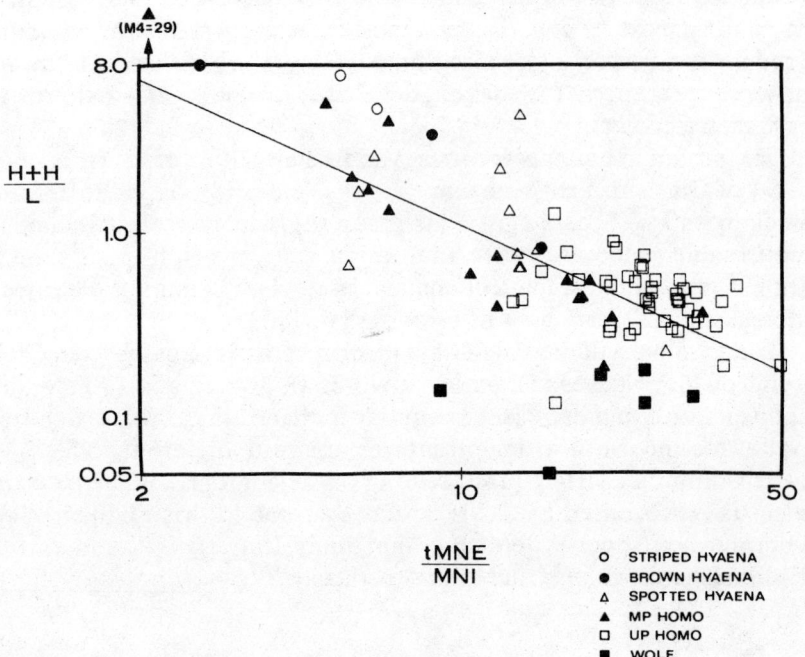

Figure 8.9. Log-log regression of anatomical completeness (tMNE/MNI) and the proportion of head and horn parts to limbs (H+H/L) for medium ungulate remains collected by human and non-human predators at shelters (from Stiner 1991b). Regression statistics in Table 8.3.

defined here simply as the ratio of cranial (head and horn/antler) elements to limbs.

Figure 8.9 shows a log-log regression of tMNE/MNI and H+H/L for 6 kinds of bone-collecting predators: three species of hyaena, Middle Paleolithic hominids, Upper Paleolithic humans and wolves (data and sources are presented in Stiner 1991b). The plot indicates a strong negative relationship between anatomical completeness for medium ungulate remains and the proportion of horns/heads to limbs therein ($r=-0.689$, $p<0.001$). In other words, the incidence of horn and/or head parts in shelter faunas increases as the amount of food carried way per feeding opportunity decreases.

The differences among predators follow a behavioral grade from scavenging to hunting, with obligate scavengers (striped and brown hyaenas) at the horn- or head-dominated ("high") end of the regression line, obligate hunters (wolves and, apparently, Upper Paleolithic humans) at the meaty end, and predators that do both (spotted hyaenas and, apparently, Mousterian hominids) occupying nearly the entire range. Because the two ends of the regression line can be related to known extremes in the foraging adaptations of modern nonhuman predators, the findings indicate a predictable and species-independent connection between the predominant foraging strategy, food transport "choices" and anatomical representation in shelters.

Table 8.3 presents regression statistics for the plot of all six predators, as well as for a variety of species subsets. All groupings display significant negative relationships between tMNE/MNI and H+H/L at the 0.001 level of probability. Assorted recombinations reveal that only the addition of the wolf group visibly affects the r^2 values, but even this set does not reduce the probability that the relationship among any of the predator sets is highly significant. The comparisons show that a broader ecological principle is at work, one that effectively cross-cuts phylogenetic boundaries between the species considered.

Figures 8.10a and 8.10b present the same data, but now the nonhuman predator cases are separated from the hominid cases, making the distributions easier to see and discuss in more detail. Some predators are confined to specific segments of the regression line, while others clearly are not. The obligate hunters show the least amount of variation. Cases produced by obligate scavengers display somewhat greater variation. Predators that do both display the most variation of all. Among the nonhuman predators (Fig. 8.10a), wolf den faunas occur only on the meaty ("low") end of the regression line. They correspond to limb-biased or complete anatomical patterns (i.e., the observed H+H/L value is roughly equivalent to the expected value) and large numbers of parts transported per carcass source (tMNE/MNI is high). Brown and striped hyaena cases tend toward the "high" end of the line, although more variation is apparent, and are usually dominated by head and/or horn elements and, on the average, contain considerably fewer parts transported per carcass source. Spotted hyaenas appear more versatile by comparison, as den cases are distributed across most of the regression line and vary in content from horn- or head-dominated patterns with low tMNE/MNI values to a relatively balanced (meaty) ratios of horn/head to limbs and higher tMNE/MNI values.

Table 8.3. Regression statistics for the relationship between H+H/L and tMNE/MNI in faunal assemblages collected by predators at shelters.[a]

predator groups	r value	r^2	N cases	p
all six predators[*]	-0.689	0.475	84	<0.001
all except MP hominids (STRH, BH, SH, UP, W)	-0.639	0.408	67	<0.001
all hominids[**] (MP, UP)	-0.707	0.500	61	<0.001
hominids and hyaenas (STRH, BH, SH, UP, MP)	-0.779	0.607	77	<0.001
hyaenas and wolves[**] (STRH, BH, SH, W)	-0.740	0.548	22	<0.001
MP hominids only	-0.873	0.762	16	<0.001

STRH - striped hyaena, BH - brown hyaena, SH - spotted hyaena, W - wolf, MP - Middle Paleolithic hominid, UP - Upper Paleolithic human.

[*] Plotted in Fig. 8.9.
[**] Plotted in Fig. 8.10.

[a] The anatomical part ratios, H+H/L (head and horn MNE divided by limb MNE) and tMNE/MNI (total MNE for all skeletal elements divided by the minimum number of individual carcass sources), are based on bone specimens, and never on teeth.

The Upper Paleolithic (UP) human cases in Fig. 8.10b are rather tightly clustered at the meaty end of the regression line, indicating little variation among cases, whereas Mousterian (MP) cases are distributed across the entire line in a manner akin to the spotted hyaena group in Fig. 8.10a. In the case of the Mousterian, however, variation is expressed within one small region of Italy, whereas variation in the spotted hyaena assemblages is expressed across several regions of Africa and Europe.

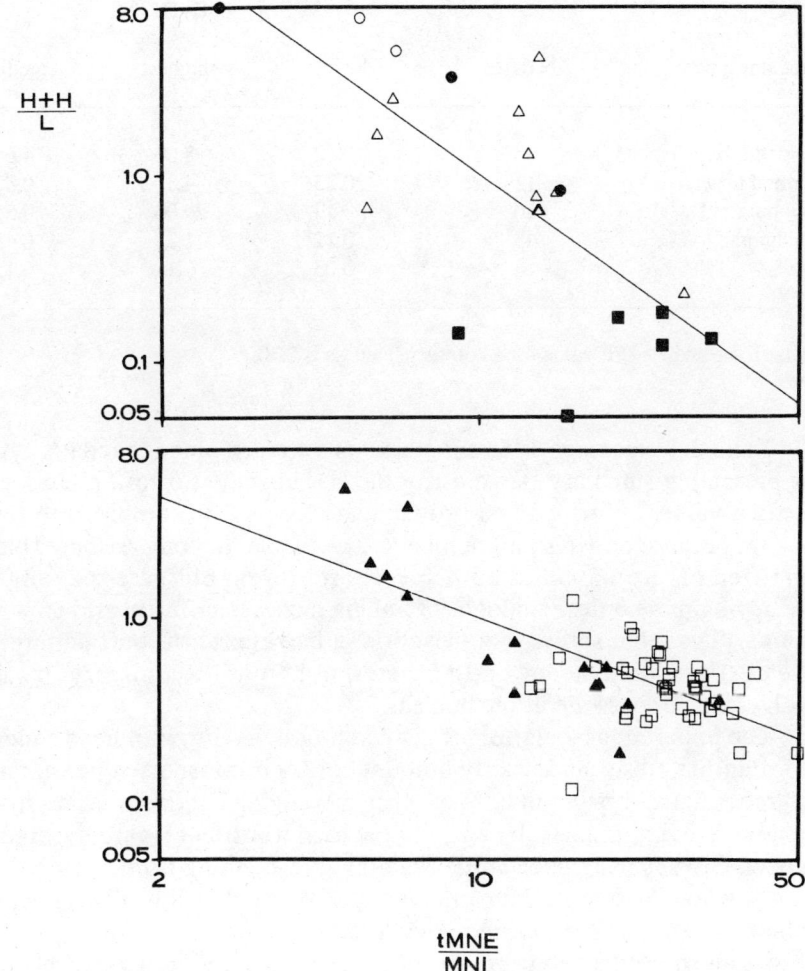

Figure 8.10. Log-log regression of tMNE/MNI and H+H/L for medium ungulate remains in shelters collected by three species of hyaena and wolves (top), and collected by Middle and Upper Paleolithic hominids (bottom, excludes M4 outlier because H+H/L=29). Symbols as in Fig. 8.9, regression statistics in Table 8.3.

Table 8.4. Summary statistics for H+H/L ratio by predator type.

predator type	N cases	H+H/L lowest	H+H/L highest	median
striped/brown hyaena	5	0.82	8.00	4.64
spotted hyaena	11	0.23	4.31	0.81
MP hominid	16	0.19	29.00	0.57
UP human	45	0.12	1.25	0.43
wolf	6	0.05	0.18	0.13

Note: Expected H+H/L value for a complete carcass is 0.30.

Table 8.4 illustrates different ranges of variation among predator types by presenting summary statistics for the H+H/L ratio (lowest, highest and median values). Striped and brown hyaena cases are combined to form one set because only a small number are available for comparison. This is justified on grounds that both species represent obligate scavengers occupying more or less analogous foraging niches in different arid environments. The table shows that variation in body part transport patterns is greatest for the hyaenas and Mousterian hominids, and very low for wolves and Upper Paleolithic humans.

The apparent association of scavenging behavior with head- and/or horn/antler-biases and scanty amounts of food transported per carcass source is not surprising in light of what scavenging opportunities normally entail. Scavenging, itself, will almost by definition heighten variation among cases, since opportunities can involve anything from a few gristly bits to whole carcasses. Hunters, on the other hand, nearly always get the whole carcass, or, if they lose it, they usually lose it all.

Modern control data involving artificial provisioning of striped hyaenas near dens (Skinner et al. 1980) show that scavengers prefer the same parts as hunters when presented with the choice (Stiner 1991b). However, naturally high levels of food dispersion characterize scavenging contexts and generally prevent such choices from being realized, except where very large prey are concerned. Probability works against scavengers getting most of a carcass most of the time, and this is why cases involving a high dependence on scavenged food will tend to be distributed on the horn/head-dominated end of the regression line.

Actualistic and wildlife data on feeding sequences of nonhuman predators at kill sites (e.g., Schaller 1967; Brain 1981; Haynes 1980, 1982; Blumenschine 1986b) also predict lower returns per carcass on the average for scavengers. This translates to fewer transport opportunities both in the number and array of parts that can be carried away. Because crania and horn/antler are among the most persistent leftovers at primary feeding sites, they are most often available to scavengers. While the sequence of carcass destruction is best known from *kill* contexts, many of the observations also apply to natural death sites. The body structure of prey is the same, and so ravaging follows the same general sequence if discovered prior to putrefication. Thus, heads are persistent remains at nonviolent death sites as well as at carnivore kill sites.

The attraction of head parts to bone-collecting predators is not simply a matter of skeletal persistence, however. Natural deaths show a strong link to starvation (Stiner 1990b, 1991b) and thus to lower nutritional values for most body parts with the notable exception of the head. Soft cranial tissues, such as the brain, display relatively high fat levels that are much more stable relative to changes in prey health than is true for any other part of the anatomy (documented in Stiner 1991b). The relative nutritional value of head parts will naturally increase in scavenging contexts and wherever fat is chronically scarce more generally, such as in many tropical environments. If heads are to be consumed by a bone-collecting predator, the incentive to move them to protected places is very high because these parts also require considerable processing. Parenthetically, hyaenas will move both horn/antler and head parts because they can effectively process and digest both, whereas hominids in Pleistocene Italy transported mainly heads to shelters (Stiner 1991a, 1991b).

In considering the results of the anatomical analysis, it is important to realize that the variation discussed here is expressed at the inter-assemblage level and overrides variation at the intra-assemblage level. The data register the degree to which foraging strategies varied in the context of bone accumulations forming over a relatively long time and in a particular place. In this way, the data reflect the behavioral *tendencies* of predators, not what they will do every situation. For example, Upper Paleolithic foragers might have scavenged from time to time, but the visible fall-out of this behavior would have been swamped by the more consistent and pervasive products of hunting at each shelter site. The Mousterian scavenging pattern stands out because the strategy was somehow spatially and temporally independent of occupations that emphasized mainly hunting. The central tendency and the range of variation in anatomical composition of transported faunas distinguishes predator adaptations and appears to correspond to ecological differences between predator niches. Variation reflects the relative emphasis on hunting versus scavenging and, apparently, the plasticity of a species in responding to varying foraging conditions. Of the predators considered, Mousterian strategies were the most varied, while Upper Paleolithic human strategies were among the least so.

PREY AGE SELECTION AND FOOD TRANSPORT AS JOINT INDICATORS OF PROCUREMENT STRATEGIES

There are few modern cases for which data on both mortality patterns and food transport are available. Both kinds of information may exist for a predator and prey species, but the two phenomena are seldom recorded for the same populations and in the same time frame. The data on modern predators summarized in previous sections nevertheless permit some general associations between prey age selection and food transport choices. For example, spotted hyaenas depend heavily on scavenged sources of food in southern African habitats and tend to produce old-biased age patterns *and* head- and/or horn-dominated anatomical patterns in den assemblages (e.g., Tilson *et al.* 1980; Henschel *et al.* 1979; Mills 1984a, 1984b). A variety of mortality patterns, ranging primarily between U-shaped and living-structure types, result from active hunting by carnivores. Among the species that collect bone in shelters, hunting generally results in "meatier", more complete anatomical patterns.

Because I collected both mortality and anatomical data for each Italian Paleolithic assemblage, it is possible to directly compare patterns of prey age selection and food transport for several of the hominid shelter occupations (Ncases=10). Figure 8.11 shows a log-log plot of anatomical representation (H+H/L) against the proportion of prime to old animals (PRIME/OLD) in each case. The graph also features two models (open squares 1 and 2), distinguished from real archaeological cases by a shaded rectangular frame. Model 1 is based on an idealized living-structure death pattern for whole prey, and model 2 is based on an idealized U-shaped death pattern for whole prey (Table 8.5). The regression is calculated without the models and indicates a strong negative relationship between

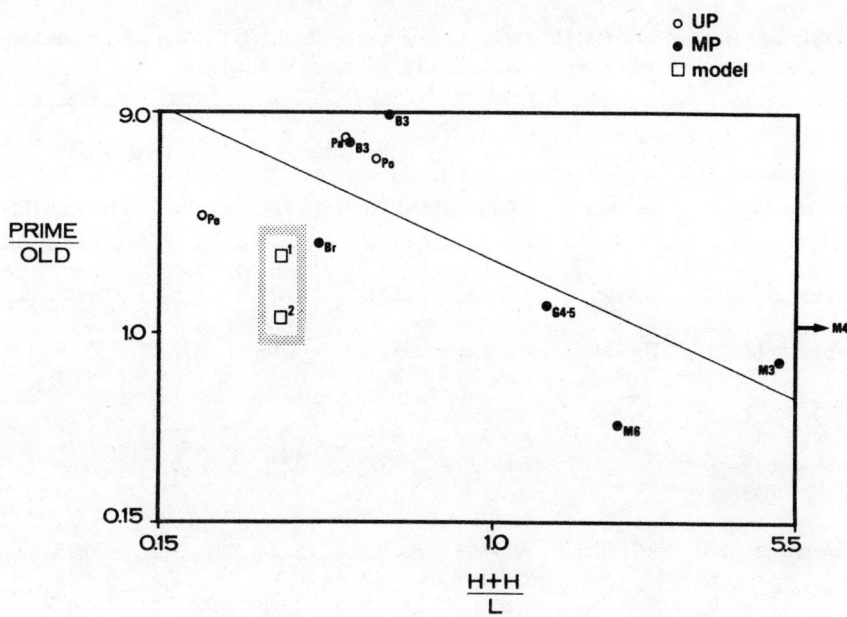

Figure 8.11. Log-log regression of the proportion of head and horn parts to limbs (H+H/L) and the proportion of prime-age to old individuals in Italian Paleolithic faunal assemblages ($r=-0.745$, $0.02>p>0.01$).

the two ratios ($r=-0.745$, $r^2=0.555$, $0.02>p>0.01$, M4 outlier excluded). This means that the proportion of head parts increases as the proportion of prime to old adults decreases; alternatively, where one finds lots of head parts and few limbs in Italian Mousterian sites, one also finds a bias toward old-aged prey.

The distribution of Mousterian cases in the plot suggests that they actually segregate into two more or less discrete clusters. Some are indistinguishable from Upper Paleolithic cases in that they show an overabundance of prime adults and relatively complete (meaty) anatomical representation. Other Mousterian cases are isolated at the "low" end of the regression line, indicating that the total Mousterian sample also incorporates a faunal complex not evident in the Upper Paleolithic sample.

Table 8.5. Head to limb (H+H/L) and prime- to old-aged adult (PRIME/OLD) ratios for medium ungulate remains in selected Italian Paleolithic shelter faunas.[a]

site/provenience	predator	prey	bone data: tMNE/MNI	bone data: H+H/L	tooth data: N	tooth data: PRIME/OLD
Palodoro (Pa)	UP Homo	red deer	13.6	0.43	5.0	7.00
Polesini (Po)	UP Homo	red deer	~49.7	~0.19	392.0	3.19
		ibex	40.0	0.51	6.5	5.67
Breuil (Br)	MP Homo	red deer	33.7	0.37	21.0	2.45
(B3)		red deer	17.0	0.55	9.0	8.80[b]
(B3)		ibex	18.0	0.44	15.0	6.69
Moscerini (M3)	MP Homo	red deer	5.1	5.00	3.5	0.75
(M4)		red deer	3.7	29.00	4.5	1.25
(M6)		red deer	5.8	2.00	4.5	0.40
Guattari (G4-5)	MP Homo	red deer	7.0	1.33	12.5	1.37
MODEL 1: Nonselective mortality, whole carcasses				0.30		2.14
MODEL 2: U-shaped mortality, whole carcasses				0.30		1.16

~ Estimated based on subset of total assemblage studied

[a] The tMNE/MNI and H+H/L ratios are based strictly on bone MNE counts, whereas the PRIME/OLD ratio is based on tooth eruption-wear data.

[b] Value has been adjusted: the divisor is actually 0, but is substituted with a value of 0.05. This procedure is a very conservative adjustment in that it brings the major outlier at one extreme closer to the average.

Note: Codes in parentheses refer to plot in Fig. 8.11.

DISCUSSION AND CONCLUSIONS

The approach described here highlights some of the boundaries of hominid foraging niches through comparisons of prey age choice and food transport by a larger array of predator species. When applied to Upper Pleistocene cave faunas in one small region of Italy, the approach demonstrates significant niche separation between Middle and Upper Paleolithic hominids. The differences correspond to changes in how hominids obtained animal prey, the extent and character of variation in their subsistence practices and, probably, the relative scales of mobility. Because considerable variation in foraging practices is indicated *within* the Middle Paleolithic as well, the findings of this study contradict some current views that Neandertal adaptations were static and inflexible. The results are discussed below, first in terms of strategic variation between cultural periods as they are currently defined, followed by consideration of increasing predatory specialization independently of the Middle-Upper Paleolithic chronological boundary.

Strategic Variation: Hunting and Scavenging

Gradation between head- or horn-dominated faunas to meaty assemblages in shelters corresponds to an emphasis on scavenging or hunting respectively. Mortality patterns provide a second means for diagnosing hunting versus scavenging at the assemblage level. Of the mortality patterns discussed, the prime-dominated and old-biased patterns are of special interest for comparative studies of hominid predatory strategies, because they are more specific to cause than most.

Perhaps the most outstanding findings on ungulate mortality and skeletal part transport in this study have to do with the *total ranges of variation* in foraging strategies for each of the two Paleolithic periods. Upper Paleolithic (and Holocene) cases cluster tightly together in both dimensions, whereas Mousterian cases are highly scattered. All hominids studied, from roughly 110,000 years ago to the present, appear to have been ambush predators, but Middle Paleolithic foragers employed a more diverse set of predatory tactics than did Upper Paleolithic peoples in Mediterranean Europe. Middle Paleolithic hominids either hunted or scavenged the very same ungulate species (especially red deer and aurochs), whereas Upper Paleolithic humans hunted them almost exclusively.

Technological evidence lends additional perspective on Mousterian adaptations. Data on tool reduction and transport in association with the faunas indicates that scavenging coincided with higher frequencies of transported and heavily resharpened tools than situations involving hunted prey, suggesting more frequent movement by hominids in the first set of circumstances (Kuhn 1990a, 1990b; see also Stiner 1990a:740-744; Kuhn and Stiner n.d.). Consistent with foraging conditions affecting modern terrestrial scavengers (Houston 1979), the archaeological data underline the importance of spatial distributions of scavenging opportunities as a major factor conditioning their procurement (see also Stiner 1990b, 1991b).

It is unlikely that any of the foraging tactics outlined above were entirely unique to either Mousterian or Upper Paleolithic hominids. However, the rules by which alternative procurement techniques were combined appear to have been quite different. Mousterian hominids were capable hunters of ungulates, for example, but they did not hunt as consistently as Upper Paleolithic foragers. Both varieties of hominid certainly were mobile, but Mousterian foraging responses may have been governed more by local, immediate exigencies than was true for Upper Paleolithic peoples (see also Kuhn 1990a, 1990b).

It is not yet clear if the level of strategic variation in food procurement within the Italian Mousterian, the character of that variation, and the geographical scale at which it occurred would be equivalent for the Middle Paleolithic of other geographical regions. In the long run, the *range* of variation may more accurately characterize Mousterian foraging patterns than the array of techniques actually used.

Increasing Predatory Specialization

Comparisons to nonhuman predators reveal that modern people are a rather specialized variety of ambush predator. From the perspective of prey age selection, the genus *Homo* appears to have evolved away from an ecologically common predatory pattern to a relatively unique one sometime in the Upper Pleistocene. Comparisons of food transport habits also suggest decreasing variation across the same time range, although these data expose no ecologically unique qualities of human foraging adaptations. The trend toward increasing predatory specialization can be considered at two separate but related levels:

1) Cases attributed to scavenging (Grotta Moscerini M2, M3, M4 & M6, and probably Grotta Guattari G4-5) comprise the earlier part of the sequence, whereas evidence of hunting is identified primarily in the later part (Grotta di Sant'Agostino and Grotta Breuil). It is conceivable at this

stage in the research that the observed variation in hunting and scavenging in the Mousterian record could represent either an evolutionary change in lifeways, *or* just seasonal or extra-annual adjustments in site use within a single, stable adaptation. The contrast between the two Paleolithic periods nevertheless is striking, even if it is not clear just when hunting became the dominant mode of foraging at every site.

2) A general trend in *hunting* practices specifically towards prime-dominated prey selection evident by the late Mousterian (Grotta Breuil), on the other hand, is difficult to deny, particularly since noticeably reduced variation persists into Upper Paleolithic through Holocene times. The Italian evidence suggests that hunting practices had become significantly more specialized by 40,000 years ago, or perhaps earlier — and within the temporal range of Mousterian technologies. The increasing focus on prime adult prey suggests a growing separation between hominid predatory niche and the niches of sympatric carnivores, and it has a number of ecological implications for understanding humans' changing place in Upper Pleistocene animal communities.

Habitual prime-dominant prey age selection is unique to humans. This class of mortality pattern diverges from natural processes of recruitment and survivorship in mammalian populations and from the typical prey selection patterns of nonhuman predators. It is the only harvesting pattern that operates *directly* upon the most reproductively active age groups at any given time, since most wild ungulate species do not begin reproducing until at least their second year of life. Prime adults also are the richest sources of stored, metabolizable fats and can be the most difficult individuals in prey populations to kill.

Prime-dominant harvesting implies considerable "selective" control on the part of hunters, either as the deliberate or inadvertent product of the strategies employed. Archaeological evidence indicates that prime-age selection cross-cuts ungulate species' ranging and social habits to a large degree (for data on other prey species, see Stiner 1990a:634-636), pointing instead to something that hunters were doing consistently and independently of the details of prey behavior (not to be confused with Hudson's findings on net-hunting of small species, this volume).

"Choice" is a sticky concept because some anthropologists are inclined to mix the consequences of a behavior with assumptions about "intention". I do not wish to make that error here; deliberate choices may have been exercised by hunters, but I assert only that human harvesting impacts prey populations in a distinct and selective way relative to broader patterns of mammalian demography.

Preferential acquisition of prime-aged prey may be facilitated by technology, particularly long-range weapons, *and/or* cooperative ambush strategies. It is important to appreciate that either of these two distinct aspects of human behavior can result in "selective" mortality patterns. The apparently great importance of cooperation among hunters in some historic situations, for example, shows that prime-dominant prey selection is not necessarily technology-bound (e.g., see review in Frison 1978b, 1984). Indeed, data on certain modern foragers suggest that extractive technology and cooperation among a larger group of hunters potentially serve as strategic substitutions for one another (Alvard and Kaplan, this volume). No doubt, long-range weapons can make the job easier, but prime-age selection also occurs within the Mousterian of Italy, where we find no convincing evidence that any kind of stone-tipped (much less propelled) projectile technology was in regular use. Even the very pronounced prime-dominant pattern of the late Mousterian is not accompanied by changes in or elaboration of extractive technologies (Stiner 1990b; Kuhn 1989, 1990a, 1990b). In the absence of such technology, prime-dominant procurement by the end of the Mousterian may indicate that relatively larger pools of human labor were sometimes available for capturing, transporting and processing ungulate prey.

Prime-Dominant Harvesting: Stable or Unstable?

Short of extermination, the potential relationships between predators and prey populations are not so much determined by how many deaths predators cause in a given year relative to other agencies as by which age groups they take. With the advent of selective ambush hunting, humans entered a new ecological relationship with their prey, and humans' impact on prey demography may have changed in significant and unprecedented ways.

Nonhuman predators inadvertently leave the most productive age cohorts free to generate more prey — the main source of new infants in the coming year. Cursorial carnivores, for example, are most likely to take young and old animals, but even ambush carnivores do not take an especially heavy toll of the prime age cohort. The low potential for affecting prey population growth by either class of predator is not surprising given the long evolutionary history of predator-prey relationships between the carnivores and the ungulates (e.g., Eisenberg 1981; Kurtén 1971). Humans, on the other hand, are unusual among predators in that they concentrate on prime adult prey in many situations. Such practices certainly have altered the structures of living populations in historic times and may have done so in the past as well (e.g., Koike and Ohtaishi 1987). That this kind of harvesting *necessarily* leads to situations of overexploitation is unlikely, however (see also Lyman 1987b).

The success of many game management programs today demonstrates that balanced predator-prey relationships are achievable using the prime-dominant exploitation scheme. However, trophy hunting, which involves controlled prime-dominant harvesting, can leave characteristic traces on the age structure of living populations in game parks. This is why census data for many modern ungulate populations indicate some deficiency in late-prime and especially aged adults relative to an idealized living-structure model (e.g., Lyman 1987b:138-140; see also review in Stiner 1990b). Graphically, these cases appear to be "compressed" toward the low end of the old-age axis in the triangular plots. The populations are viable, but exhibit a decreased mean lifespan and, in some cases, a lower mean age of first reproduction (especially in males). The populations may just remain in a "growth" phase rather than achieving the classic stable structure of populations unaffected by humans. Modern examples of the latter are quite rare, but may still be found in situations where hunting by humans is either not allowed or not practical (e.g., Mohave feral burros, Johnson et al. 1987).

Like modern trophy-hunted death assemblages, the archaeological cases dating to after about 40,000 B.P. are also compressed toward the low end of the old-age axis. The archaeological cases fit within the expected range of variation in the continuum between living-structure and prime-dominant mortality pattern domains, but often are deficient in late prime and aged prey, a situation not observed in death assemblages generated by nonhuman predators.

It is impossible to know the real composition of ancient prey populations, but it is very interesting that procurement practices of *late* Mousterian, Upper Paleolithic, Paleoindian, historic Eskimo, and other traditional human cultures resulted in mortality patterns equivalent to those arising from managed trophy hunting in modern game parks. The mechanisms preventing overexploitation in the past certainly differed from those of the modern era, but the net effect on prey populations appears much the same in terms of age selection. When archaeologists find evidence of compressed age structures in death assemblages associated with prehistoric human activities, what do these findings really mean with regard to humans' long-term ecological relationships with their prey? There can be no doubt that humans have hunted some species to extinction, particularly in the last few centuries (e.g., Diamond 1989). It nevertheless is essential to guard against equating what may be basic characteristics of human predatory niche with the results of intense short-term exploitation of a natural resource at the local level. People may engage in the second phenomenon frequently and as a matter of course — they move on to other resource patches (or areas) when the current one is depleted — without overstepping the dynamic balance at the community level.

The repetition of "compressed" age structures in prey death populations in archaeological records may simply reflect a basic property of the human predatory niche that has been around at least as long as anatomically modern humans in Europe and probably longer. The great antiquity of these practices shows that there is no *a priori* reason to assume that this unique predator-prey relationship was unstable. The balance required for prime-focused exploitation may have been more delicate and the rules of regulation somewhat different than those typical of nonhuman predators, but its 40,000+ year history suggests considerable ecological stability. Something else, such as habitat reduction and/or human overpopulation because subsistence is based on other sources of food, is required to push it over the edge.

Where Evolutionary and Ecological Processes Meet

Archaeologists experience much difficulty in agreeing upon appropriate circumstances, scales of observation and effective means for demonstrating that specific patterns in the archaeological record are of evolutionary significance. One of the impediments to investigating evolutionary issues from the perspective of subsistence has been archaeology's traditionally exclusive focus on humans. Like every species on earth, humans are unique. Yet an interspecific perspective relating characteristics of faunal accumulations to basic kinds of behavior proves very useful for addressing questions about human lifeways and humans' ecological position among predators. The practice of researching key variables independently of the human situation represents one effective way to maximize the informative value of the archaeological data.

ACKNOWLEDGMENTS

I am grateful to Lewis R. Binford, Steven L. Kuhn, Erik Trinkaus, Lawrence G. Straus and Diane Gifford-Gonzalez for their guidance and critical input throughout the course of this research. Robert D. Leonard,

R. Lee Lyman and Richard G. Klein have also provided constructive critiques for the final synthesis. I thank Lew Binford and Gary Haynes for access to unpublished material and notes on modern control faunas, and my European colleagues, A. G. Segre, E. Segre-Naldini, A. Bietti, P. Cassoli, D. Cocchi, A. Radmilli, C. Tozzi, and G. Manzi, for their assistance and encouragement during data collection in Italy. This research was supported by the American Association of University Women, the L. S. B. Leakey Foundation, the National Science Foundation (dissertation improvement grant, BNS-8618410) and the Institute for International Education (Fulbright Program).

NOTES

1. The two Saskatchewan wolf cases in the aberrant "cursorial" group are somewhat biased toward prime adult prey because ambush behaviors took place at caribou calving grounds. Pregnant females temporarily aggregated at this location in early spring, and the two cases represent series of kills made before and just after the birthing peak respectively (Miller 1975).

2. Because the comparisons involve so many mammalian species, Latin names are omitted from the text to enhance the flow of the presentation.

9

Subsistence Change and Pinniped Hunting

R. Lee Lyman

INTRODUCTION

Analysis of demographic (age and sex) data derived from zooarchaeological remains of prey species has, over the years, resulted in a variety of interpretations concerning whether prey populations were wild or domesticated, randomly or selectively harvested (e.g., Elder 1965; Wilkinson 1976), or hunted as opposed to scavenged (Klein 1982a). These kinds of data have also been used to investigate the type of hunting techniques used (Emerson 1980; Levine 1983), predator-prey relationships (Smith 1974), and intensity of hunting pressure in prehistory (Koike and Ohtaishi 1985). However, many such analyses have ignored animal behavior, especially variation in escape behaviors across different age-sex groups (but see Simmons and Ilany 1977 and Wilkinson 1975 for two important exceptions), instead simply comparing archaeologically observed kill demography with modern census data for an extant population of the taxon under study in order to infer human subsistence behaviors. While theoretically a reliable and independent source of reference, modern census data may be derived any number of ways. This raises the possibility that they are biased, particularly if the methods are not clearly stated or if the censusing methods are inappropriate relative to the species sampled (see also Chapter 1). Alternatively, census data may provide an accurate estimate of a population's demographic structure yet inaccurately characterize the exploited prehistoric population (see discussions in Coe 1980; Coe *et al.*

1980; Western 1980). The results of such comparative analyses might be subject to dispute if more than one set of modern population data are used (but see Cribb 1987 for some innovative suggestions on how to avoid some of these problems.)

The behavioral interactions between prey species of different age-sex categories and predators can impact the resulting age and sex structure of the death population. This is an especially important consideration in the case of humans whose hunting behaviors are at least in part dictated by available technology and an extensive knowledge of prey species behaviors. An opportunistic hunting tactic that is nonselective with regards to prey age and sex, for example, *may or may not be* independent of the technology employed. Some food-getting technologies, such as various traps, will result in the procurement of prey items independent of the age-sex (or size) of individual prey. Other technologies, various ones which take individual prey items one at a time, will result in the procurement of only certain categories of age-sex (or size) prey being sought. These are not trivial distinctions, particularly when inferences of human selectivity in the taking of prey are desired. Some control of prey item procurement technology is mandatory to such inferences (see also Chapters 5, 6 and 8).

In this chapter I explore two issues. The first concerns the varying behavioral patterns of prey on land as opposed to in the sea and their potential impact on the age and sex (and sometimes size) structure of death populations in archaeological contexts. Second, I offer an explanation for an apparent change in subsistence, based on demographic characteristics of the archaeological death populations. Both parts of the investigation are accomplished without detailed consideration of modern census data, partly because of the potential biasing factors outlined above and, more importantly, because such data are poor for many of the aquatic mammals under study. Thus I attempt to investigate patterns of human exploitation in the absence of reliable information on broad population composition, instead focusing upon known prey social and reproductive behaviors in conjunction with technological data from the archaeological record. A brief review of the artifacts associated with the faunal remains suggests the procurement technology is not significantly affecting the demography of the killed population.

METHODS AND MATERIALS

Here I examine demographic data for faunal remains recovered from three late Holocene sites on the Oregon coast. All three sites contain shell midden deposits of varying extent, and dating between 3000 and 200 years B.P. (see Lyman 1988, 1989). In keeping with previous reports on these remains, I refer to the assemblages in terms of stratigraphy and radiocarbon dating. The Umpqua/Eden site yields three of the assemblages considered here: UEI, dated between 3000 and 2000 B.P.; UEII, dated between 2000 and 1000 B.P.; and UEIII, dated between 900 and 200 B.P. The Seal Rock site consists of one assemblage, SRI, dated between 400 and 100 B.P. Three assemblages from the Whale Cove site are also considered: WCI, dated between 3000 and 2800 B.P.; WCII, dated between about 2500 and 600 B.P.; and WCIII, dated between 500 and 200 B.P. I also discuss a faunal assemblage from a fourth site, Yaquina Head, reported by Minor et al. (1987) and dating between 3300 and 2000 B.P. While no demographic data are available for the mammalian remains from this site, associated materials help identify subsistence practices during the earliest known human occupation of the Oregon Coast.

Most demographic analyses of zooarchaeological remains have, to date, concerned mollusks or terrestrial herbivores. Here I focus on seven mammalian taxa. These include two artiodactyl species, deer (*Odocoileus* sp.) and wapiti (*Cervus elaphus*), whereas the remainder are marine carnivores. Of these, one is the sea otter (*Enhydra lutris*), and the others are pinnipeds (seals and sea lions), including the harbor seal (*Phoca vitulina*), northern fur seal (*Callorhinus ursinus*), northern or Steller's sea lion (*Eumetopias jubatus*), and the California sea lion (*Zalophus californianus*). I am particularly concerned with the age and sex composition of the death populations of harbor seals and northern sea lions, as these two taxa are abundant and vary considerably in kill demography and escape behaviors.

In their recent book *The Archaeology of Prehistoric Coastlines*, Bailey and Parkington (1988:6) repeat a common conception regarding the prehistoric exploitation of sea mammals. Many such taxa, they write, "require a degree of technological ingenuity and social organization which could well have been beyond the intellectual capability of [pre-Holocene] human populations; or else demand a cost in terms of physical danger which might have been a chronic disincentive until a very late period in prehistory." This view that a complex technology is necessary for exploiting pinnipeds is common among archaeologists working on coastal sites (e.g., Borden 1975; Carlson 1979; Clark 1946; Conover 1978; Hildebrandt 1984; Jobson and Hildebrandt 1980; Minor et al. 1987; Renouf 1988). It stems in part from a basic assumption that prey usually were taken directly from the sea — a demanding situation both in terms of technology and human cooperation — rather than from shoreline areas used for reproductive and other purposes. I believe that a complex technology, extensive cooperation among hunters and complex social organization were *not* prerequisites for taking most Oregon coast pinnipeds. Rather, only clubs, perhaps but not necessarily harpoons, and only one or two hunters working together could exploit a seal rookery effectively (Lyman 1989; Lyman *et al.* 1988). This hypothesis is based on information about the behavioral patterns of those taxa during the breeding season (see also Yesner 1987) and ethnographic data on pinniped hunting along the adjacent northern California coast. Ethnographic data indicate that (1) complex technology and human social organization are unnecessary for pinniped hunting and (2) the hunting methods used were both labor-efficient and safe. It is significant that these practices most often occurred during the late spring-early summer at rookeries (Kroeber and Barrett 1960).

Age-sex data for pinniped remains are of critical importance for determining whether prehistoric hunters exploited *rookeries* (terrestrial breeding areas) or *haul-outs* (terrestrial resting areas). In general, exploitation of rookeries results in high relative frequencies of adult breeding males, near-term fetuses and newborns. Exploitation of haul-outs will produce a somewhat more random or "even" sample of age-sex classes, although perhaps skewed toward (unwary) juveniles of both sexes. These general expectations will vary, as will become clear, among taxa, depending on their particular escape behaviors.

I (Lyman 1989:73) expect that "*collectors* will concentrate their efforts on a few densely occurring resources and devote minimal effort on more dispersed and less abundant resources. *Foragers* will take an equally broad or broader range of resources than pursuers, but in more equal abundances as resources are more or less evenly dispersed across spatio-temporal contexts" (Fig. 9.1). These expectations employ the collector-forager distinction first outlined by Binford (1980). Here, I infer a collector strategy if the relative abundance of one of the taxa considered equals or exceeds the average relative abundance of all taxa plus two standard deviations (SDs). If the relative abundance of one or more taxa exceeds the mean plus one SD, I infer a combined collector-forager strategy. If no taxon's relative abundance exceeds the mean plus one SD, I infer a forager strategy (Lyman 1989). Because the collector-forager dichotomy, as I conceive of it, denotes variation in resource acquisition, and not resource returns, I utilize only the numbers of identified specimens (NISP) as measures of taxonomic abundances. Other researchers have used biomass of prey (Hildebrandt 1984; Yesner 1988), a variable that may confuse resource acquisition activities with resources acquired. The biomass variable might be more appropriate when the analyst seeks some

Subsistence Change and Pinniped Hunting

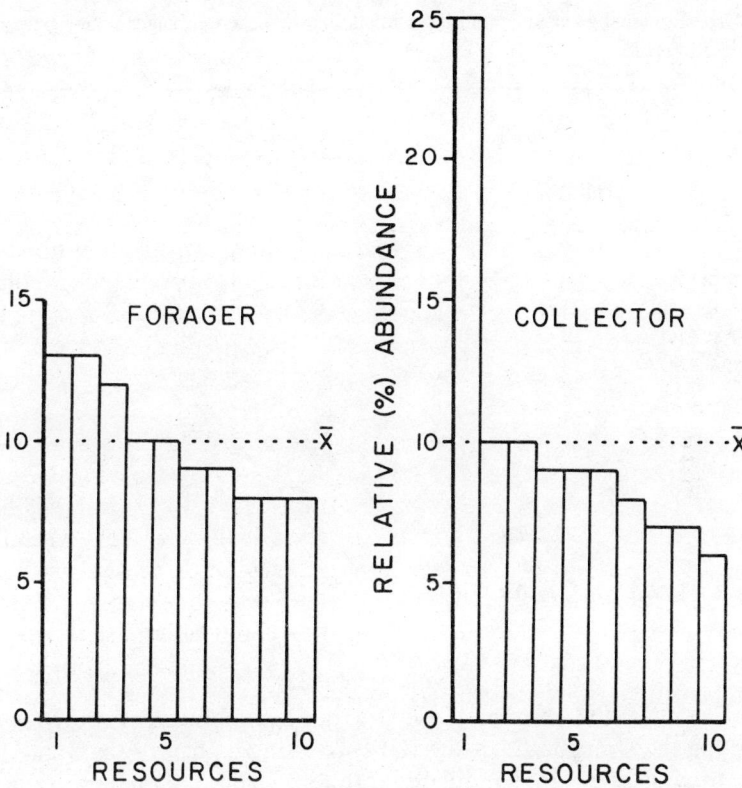

Figure 9.1. Modeled variation in the relative proportions of resources exploited by foragers and collectors.

measure of costs and benefits, as in cases where optimal foraging theory takes an explanatory role. In this study, however, I am interested in resource acquisition specifically.

RESULTS

Collectors or Foragers?

Results for the three assemblages greater than 2000 years old (Table 9.1a; Fig. 9.2) suggest a collector strategy focusing on wapiti (and deer, secondarily) at the Whale Cove site (WCI) and on deer (and wapiti to a

Table 9.1a. Frequencies of seven mammalian taxa in three assemblages dating between 3300 and 2000 B.P.

	WCI		Yaquina Head		UEI	
taxon	NISP	%	NISP	%	NISP	%
harbor seal	9	4.0	3	3.7	13	40.6
northern fur seal	21	9.3	8	9.9	0	0.0
Steller's sea lion	18	8.0	6	7.4	1	3.1
California sea lion	4	1.8	2	2.5	0	0.0
sea otter	7	3.1	14	17.3	5	15.6
deer	51	22.7	28	34.6	8	25.0
wapiti	115	51.1	20	24.7	5	15.6
total	225		81		32	
average%		14.3		14.3		14.3
SD%		16.4		11.0		13.9
average% + SD%		30.7		25.3		28.2

lesser extent) at the Yaquina Head site, but a combined collector-forager strategy at the Umpqua/Eden site (UEI) focusing on harbor seals and deer. These three assemblages indicate large terrestrial mammals were being acquired in abundances roughly equal to or greater than marine mammals prior to 2000 B.P.

Results for all five assemblages younger than 2000 B.P. (Table 9.1b; Fig. 9.2) indicate collector strategies. Of these, four assemblages indicate a focus on harbor seals (UEII, UEIII, WCII, WCIII), and the fifth assemblage indicates a focus on Steller's sea lions (SRI). Cervids tend to rank second, third, and/or fourth in abundance, indicating they were being acquired less frequently after 2000 B.P. than in earlier times.

The data suggest a shift towards more intensive collector-type subsistence pursuits. As shown in Fig. 9.2, relative taxonomic abundances in the assemblages dating to before 2000 B.P. are more even than those in the five post-2000 B.P. assemblages. In other words, prey species selection by hunters appears to have been more generalized prior to about 2000 B.P. Later assemblages indicate relatively more specialized acquisition focusing primarily on one mammalian species, probably a locally concentrated resource. Analyses of skeletal part data indicate no significant differences

Subsistence Change and Pinniped Hunting

Table 9.1b. Frequencies of seven mammalian taxa in five assemblages dating between 2000 and 200 B.P.

taxon	UEII NISP	%	UEIII NISP	%	SRI NISP	%	WCII NISP	%	WCIII NISP	%
harbor seal	146	73.7	269	53.2	33	1.8	36	69.2	39	81.2
n. fur seal	14	7.1	17	3.4	97	5.4	0	0.0	0	0.0
S. sea lion	3	1.5	23	4.6	984	54.8	1	1.9	2	4.2
C. sea lion	2	1.0	7	1.4	31	1.7	0	0.0	0	0.0
sea otter	17	8.6	55	10.9	141	7.8	1	1.9	2	4.2
deer	10	5.1	66	13.1	262	14.6	8	15.4	4	8.3
watpiti	6	3.0	68	13.4	249	13.9	6	11.5	1	2.1
total	198		505		1797		52		48	
average%		14.3		14.3		14.3		14.3		14.3
SD%		24.4		16.5		17.2		23.1		27.4
average% + 2SD%		63.1		47.3		48.7		60.5		69.1

in representation of pinniped carcass body parts, indicating that whole animals were brought to the site for processing. Intensification of harbor seal exploitation from UEI to UEII times is apparent at the Umpqua/Eden site, after which exploitation decreased in intensity by UEIII times. At the Whale Cove site, a shift from cervids to an increasing focus on harbor seals occurred from WCII to WCIII. Why did exploitation of harbor seals intensify at Whale Cove over three temporal periods, while initially intense exploitation of seals declined at Umpqua/Eden?

Technological comparisons of the frequencies of harpoon parts and arrow heads in the sites show that harpoon parts are rare in all but SRI (Table 9.2), where resource acquisition focused on adult male Steller's sea lions (see also Lyman *et al.* 1988). Artifact samples for assemblages predating 1000 B.P. are too small to place any trust in relative frequencies. Note, however, that the ratio of harpoon parts to arrow heads decreases from 0.25:1 in UEII to 0.13:1 in UEIII. While not entirely true, arrow heads are generally equated with terrestrial game procurement. Thus this single change may signify a subsistence shift back to terrestrial mammals (cervids), as is also supported by the faunal data (Fig. 9.2). However, it is

194

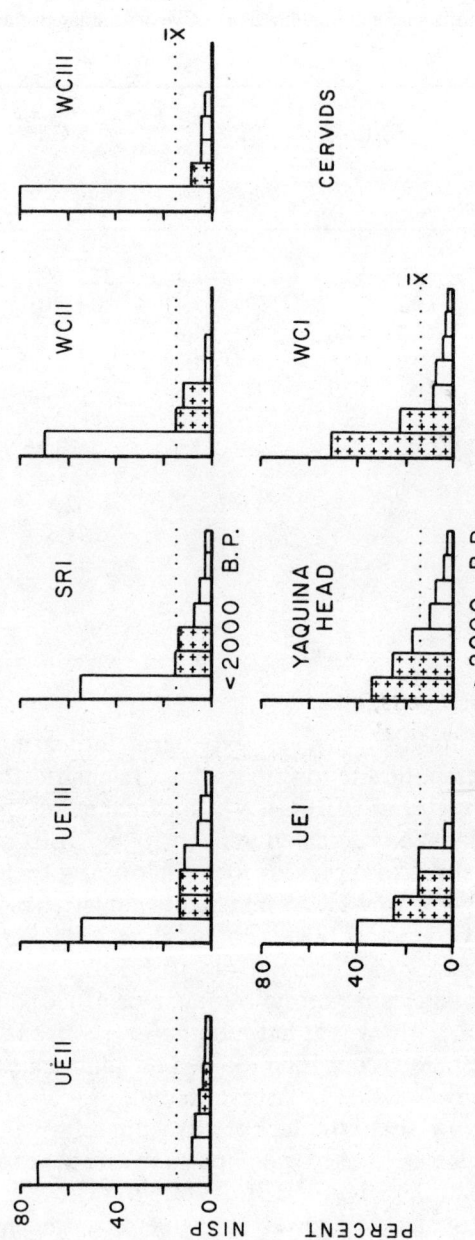

Figure 9.2. Observed variation in the relative proportions of seven exploited mammalian taxa for three Oregon coast assemblages greater than 2000 years old and five assemblages less than 2000 years old (data from Table 9.1). Shaded bars denote cervids. Note that later assemblages display a more collector-like frequency distribution than earlier assemblages (compare with Fig. 9.1).

Table 9.2. Frequencies of harpoon parts and arrow heads associated with Oregon Coast faunal assemblages.

assemblage	number of harpoon parts	number of arrow heads
SRI	34	38
WCII-WCIII	0	5
UEIII	6	46
UEII	1	4
UEI	0	0
WCI	0	2
Yaquina Head	0	2

important to acknowledge that arrows could also have been used to take pinnipeds while on land, especially on rookeries, as is demonstrated by an recent photograph of a man narcotizing a male California sea lion with a bow-and-arrow (Peterson and Bartholomew 1967:Fig. 4). Ethnographic data also reveal that wooden clubs measuring about 2 meters in length were used to take pinnipeds, although such artifacts are unknown in the Oregon coast archaeological record. Their absence from archaeological sites may be explained any number of ways: such expedient or impromptu tools could have been discarded at the kill site, or simply not preserved in the sites sampled thus far. Three whale bone clubs, each about 0.8 meters long, are known from the same record (Lyman 1989:85).

The technology known both from ethnographic accounts and archaeological sites would have sufficed in on-shore (arrows, clubs, harpoons) or off-shore (harpoons) hunting contexts. However, there is no indication that prehistoric hunters (in so far as those can be measured by artifacts) created the observed age-sex classes of pinnipeds in archaeofaunas by procuring them from distant offshore contexts (see Jobson and Hildebrandt 1980; Hildebrandt 1984; Lyman et al. 1988 for discussion of this issue). Variation in age-sex classes can be explained by species-specific social adaptations of pinnipeds and their escape behaviors when threatened on land.

Hunting Pressure

R. G. Matson (1983) suggested several years ago that over-hunting of pinnipeds in littoral settings using a relatively simple technology is more likely to occur than over-hunting with a complex marine-oriented technology in pelagic settings. Given that about 83% (145 of 174) of the ageable harbor seal remains from Umpqua/Eden are of newborns, it seems likely that human inhabitants of that site exploited nearby estuarine rookeries with what I suspect was a relatively simple hunting technology (see above and Lyman 1989; Lyman et al. 1988).

Harbor seal pups less than six months of age are not very wary (Bigg 1969), whereas older individuals are extremely so and easily scared away from haul-outs (Mate 1981; Orr 1972). Pups are born on land, but the birthing area is often located below the high tide line and they can swim, if only weakly, within hours after birth (Banfield 1974; Bigg 1969; Newby 1973, 1978). Pups spend about half of their time on land thereafter (Lawson and Renouf 1987). While mothers will protect their pups on land, they do not do so very aggressively and will abandon pups when seriously threatened (Lawson and Renouf 1987; Newby 1978). It is significant that mothers and newborns form nursery groups isolated from the main colony, and youngsters spend much time sleeping and playing within this restricted area (Newby 1973, 1978). Some prenatal and neonatal mortality occurs in this context (Newby 1973), and about 15% to 20% of the newborns do not survive to the age of one year (Bonner 1979). While males fight over females in estrus, fighting is not as violent as that observed in other pinniped taxa (Bigg 1969).

A low rate of mortality in harbor seal pups and juveniles is in no way adverse to population stability in most situations. However, pups can be easily preyed upon by humans armed with relatively simple weapons while on shore, and the principal problem limiting exploitation has to do with how many are killed and how often a particular rookery is harvested. Hunting of the estuary population of harbor seals near Umpqua/Eden prior to about 2000 B.P. may have been relatively more sporadic and less intensive (only four specimens of newborns and no ageable specimens of older individuals are associated with UEI). After 2000 B.P. exploitation intensified and may have reduced the population, particularly by about 1000 B.P., to such a point that alternative resources had to be found (84% of ageable remains in UEII are newborns; 82% of ageable remains in UEIII are newborns). The possibility that intensive hunting can affect the viability of harbor seal populations is well documented for the late 19th and early 20th centuries, when many pinnipeds were killed for bounties (e.g., Pearson and Verts 1970). The faunal data shown in Table 9.1 imply that alternative resources exploited by human occupants of Umpqua/Eden included cervids.

Steller's sea lion remains from the Seal Rock site require a different explanation. This species utilizes the same rookeries year after year, breeding and pupping mainly in June plus or minus two or three weeks (Orr 1972; Orr and Poulter 1967). Newborns frequently drown, but by the age of three months they have become adept swimmers (Gentry and Withrow 1978). Dominant males do not leave their territory during the breeding season, even when threatened, whereas other age-sex classes dash into the ocean when disturbed (Orr and Poulter 1967). Breeding males fight violently, sometimes crushing pups. In other cases, mothers abandon pups and they die of starvation (Gentry and Withrow 1978; Orr and Poulter 1967). Mortality rates are largely unknown for Steller's sea lion juveniles and adults; pup mortality has been observed to vary from 10 to 100% between years (Mate and Gentry 1979). The northern elephant seal (*Mirounga angustirostris*) is another large, violent-fighting species and is quite gregarious. Studies of rookeries indicate five major factors contribute to elephant seal pup mortality. These are the density of the rookery population per unit area occupied, female aggression towards unrelated young attempting to nurse, crushing by fighting males, severity of storms, and mother-infant separation (LeBoeuf and Briggs 1977). The last four are in fact dependent on the first to a significant extent, as increasing population density restricts pup movement and enhances the probability that one or more of the other factors will also take effect. Thus, reducing the population at rookeries, especially by culling breeding-age elephant seal males, could actually have a positive effect by decreasing density and thereby reducing one of the most serious direct threats to pup survival — crushing by fighting males. The supply of substitute males ready to form harems is in no way threatened, as long as breeding-age males are not over-hunted. Similar culling patterns of Steller's sea lion adult males from the rookery at Seal Rock would also be expected to enhance survival of juveniles in the population within the limits defined here. Seventy-five percent of all Steller's sea lion remains (745 of 948) recovered from Seal Rock are adult, breeding-age males, 16.6% (NISP=163) are of adult females, and only 5.8% (NISP=57) are of near-term fetuses or newborns (19 specimens could not be assigned to age-sex class) (Lyman 1989). These values suggest adult male Steller's sea lions were the main targets of pinniped hunters from Seal Rock. This age-sex class is the one most susceptible to predation by terrestrially-bound human hunters armed with clubs and harpoons (Lyman 1989).

Finally, the Whale Cove site data suggest that exploitation of harbor seals intensified through time. Only 10 of the 17 (58.8%) total ageable harbor seal remains represent newborns. Unlike the situation at Umpqua/ Eden where over 80% of the harbor seals were newborns, hunters from Whale Cove were not exploiting a rookery, or, if they were, exploitation was of much lower intensity than that evident at Umpqua/Eden. For this reason, human exploitation of adults would have had less impact, I suspect, on the Whale Cove population's survival, thereby allowing some intensification without adverse impacts to the local population. A more random selection of all age-sex classes, if not practiced too intensively, would reduce intraspecific competition for food and thus enhance survival of young animals in the population, thereby evening-out the effects of predation. Youngsters would have a better chance of surviving to adulthood due to reduced intraspecific competition with adults for food. I emphasize, however, that my inferences are largely conjectural for this site, because the sample of harbor seal remains from this site is small, making it difficult to track temporal changes in the exploitation of different age-sex groups through time.

CONCLUSIONS

Hunters on the Oregon coast prior to 2000 B.P. tended to emphasize terrestrial mammals more than marine mammals, and their subsistence pursuits more closely resemble foraging rather than collecting as modeled here. After that time, hunters apparently shifted to a more collector-like strategy aimed at exploiting pinniped rookeries. It is not clear why the shift occurred, although clarification of some of its manifestations should aid in isolating the causes through further research based on a larger array of cases. The shift may relate to a decrease in residential mobility, as indicated by an increase in the number of substantial structures (houses) at sites around this time. The eventual return to an emphasis on cervids at Umpqua/Eden by around 1000 B.P. may have been a response to overhunting of local pinniped populations, as also suggested by the shifting ratios of harpoon parts and arrow heads.

While Matson's (1983) overkill hypothesis can be made to account for the Oregon coast pinniped remains when age-sex data and pinniped behavior are considered, I hasten to note that we know much less about the effects of hunting and selective culling of particular age-sex classes from pinniped populations than we do about terrestrial mammals (e.g., Lyman 1987b and Yesner 1988 and references therein). The few studies available (e.g., York and Hartley 1980) indicate such hunting will modify infant/ juvenile survivorship, and thus population productivity between years. The scenarios I have offered should therefore be viewed as hypotheses requiring additional analyses and data to corroborate them.

ACKNOWLEDGMENTS

Study of the Oregon coast faunal collections was made possible by a grant from Oregon Sea Grant (Grant NA85AA-D-SG095, Project R/CP-25). Oregon Sea Grant is partly funded by NOAA, Office of Sea Grant, Department of Commerce. I thank K. M. Ames, A. C. Bennett, D. R. Brauner, L. A. Clark, R. E. Ross and D. N. Schmitt for their help during my research on those materials. Thanks also to Mary Stiner for convincing me of the value of this paper.

10

Thule Eskimo Subsistence and Bowhead Whale Procurement

James M. Savelle and Allen P. McCartney

INTRODUCTION

Investigations of the relationships between prehistoric hunter-gatherer societies and very large mammals (megafauna) have had a long history in archaeology. Most such investigations have traditionally centered on human-megafauna relationships as they relate to extinct terrestrial taxa such as mammoths and mastodons (e.g., Martin 1967; Saunders 1980; Soffer 1985; Haynes 1985; Frison and Todd 1986; Fisher 1987; Klein 1987). The primary questions typically concern distinguishing between (a) natural and cultural deposits, (b) if cultural, active hunting as opposed to scavenging, or (c) if active hunting can be demonstrated, opportunistic versus specialized "megafauna hunting" strategies as reflected by patterns of age/sex selection. Although receiving far less attention, the investigation of large marine mammal hunting/scavenging by more recent prehistoric societies also concerns many of the same questions. This paper discusses recent research on prehistoric Thule Eskimo subsistence practices (ca. A.D. 1000-1600) of the eastern Canadian Arctic, focusing on the nature and extent of bowhead whale use through the examination of mortality data. Bowhead whales are the largest prey species of any known prehistoric or historic hunter-gatherer society, and exploitation of this prey species was potentially a major factor determining Thule subsistence and settlement patterns.

THE BOWHEAD WHALE AS A PREY SPECIES

The bowhead whale (*Balaena mysticetus*) is unique in the central Arctic as a prey species. It is by far the largest of arctic mammals, with mature adults attaining lengths of up to 20 m (Nerini *et al.* 1984) and weights up to 100 tons (Reeves and Leatherwood 1985:306-308). Bowheads are adapted to ice-margin zones, and follow the advance and retreat of the ice on a seasonal basis (Marquette 1978; Reeves *et al.* 1983; Reeves and Leatherwood 1985). Accordingly, they are a relatively predictable seasonal resource during the short open water season of central Arctic environments. Bowheads are slow swimmers, averaging 2-4 knots, feed near the surface, are relatively easy to approach, and float remarkably well after death (McVay 1973; Marquette 1978; Reeves *et al.* 1983). Finally, the very large size of their bones (Fig. 10.1) make them ideal for constructing dwellings in an area lacking trees or significant quantities of driftwood.

Four major geographic stocks of bowheads are recognized: Spitzbergen, Davis Strait, Bering Sea, and Okhotsk Sea (Mitchell 1977). The Davis Strait stock, from which Thule Eskimo in the eastern Canadian Arctic obtained individuals, is estimated to have numbered approximately 11,000 as late as 1825, following over 100 years of whaling by Europeans (Mitchell and Reeves 1981). Commercial whaling of this stock ended in approximately 1915, as a result of severe stock depletion. There is little evidence that the stock has recovered, and recent estimates place it at only a few hundred (Davis and Koski 1980).

PREVIOUS INTERPRETATIONS OF THULE WHALING

Characteristic of many prehistoric Thule Eskimo sites in the eastern Canadian Arctic are semisubterranean winter dwellings in which bowhead whale bones constitute a primary structural material (e.g., roof pillars and rafters, and wall supports). These bones often occur in impressive quantities, and sites of 15-25 dwellings may contain the remains of over 100 individual whales (see e.g., McCartney 1978, 1979). On the basis of these structural components, Mathiassen (1927:85), who originally described and defined the Thule culture, suggested that "Whaling has apparently been one of the principal occupations." Subsequent researchers, while recognizing both "whaling" and "non-whaling" regional variants, explicitly or implicitly followed this reasoning: the presence of whale bones as

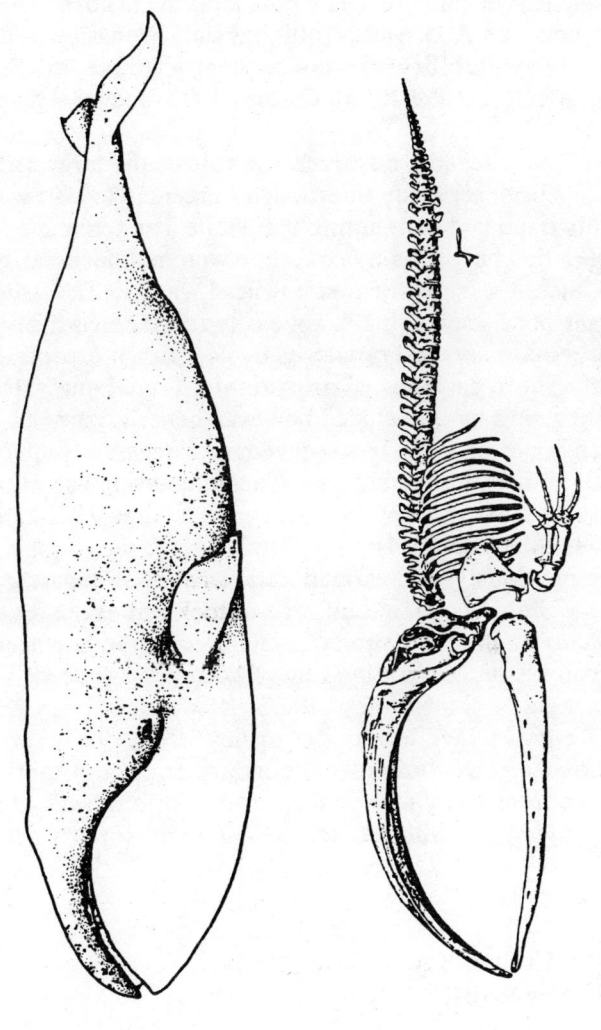

Figure 10.1. Bowhead whale (*Balaena mysticetus*) and skeleton (from McCartney 1980).

dwelling construction materials implied active whale hunting. The initial development of Thule culture in Alaska prior to ca. A.D. 1000, the migration of Thule Eskimos across northern Canada and into Greenland, and the subsequent 'decline' of Thule populations and abandonment of the High Arctic between A.D. 1400-1600 have all been related directly to bowhead whale availability, and consequently, active bowhead whale hunting (e.g., McGhee 1969/70; McCartney 1977; and review in Maxwell 1985).

Within the past decade, however, the role of the bowhead in Thule subsistence has been seriously questioned. Freeman (1979) was the first to address this issue in detail, noting that while Thule Eskimos were certainly *utilizing* bowhead whale *bone*, they were not necessarily actively *hunting* and, therefore, subsisting on bowhead whales. The issue of active hunting versus bone scavenging is not easily resolved, as many researchers since Freeman have acknowledged. Recent interpretations have become much more cautious. For example, faunal analysts of Thule winter dwelling sites have excluded bowhead bones because of the uncertainty of such bones deriving from scavenged carcasses as opposed to actively hunted animals and/or structural/manufacturing use as opposed to dietary use (e.g., Staab 1979; Taylor and McGhee 1979; Rick 1980; Morrison 1983; McGhee 1984).

We have previously summarized various lines of zooarchaeological, ecological, technological, social, and settlement evidence which we believe support an interpretation of extensive active bowhead whaling rather than carcass scavenging by Thule Eskimos in whale-rich regions of summer open water (McCartney 1980b, 1984; McCartney and Savelle 1985; Savelle 1987; Savelle and McCartney 1988, 1990). Much of this evidence, however, is admittedly secondary or circumstantial. In this paper, we turn specifically to mortality profiles of bowhead whales represented at Thule sites, which offer a robust measure of active whale hunting.

MORTALITY STUDIES AND HUNTING VERSUS SCAVENGING

Mortality profiles of fauna represented in prehistoric bone assemblages have received considerable attention in recent archaeological literature. Klein and Cruz-Uribe (1984), Klein (1987), as well as paleontologists such as Kurtén (1953) and Voorhies (1969) have drawn upon two basic theoretical models of age structure developed by population biologists.

These are the living-structure (also called catastrophic by some archaeologists) and the attritional death models, both of which assume stable population size and structure (see Chapter 1).

In the living-structure (catastrophic) model, successively older age categories contain proportionately fewer animals, in accordance with the age structure of the parent living population. Mortality profiles of this type may arise from "catastrophic" natural events, such as flash floods, "red tide", volcanic eruptions, and certain epidemic diseases, or, in the context of prehistoric nonselective hunting techniques that harvest age groups in direct proportion to their natural abundance, such as communal drives.

In the attritional model, "prime" age (reproductively active) adults are underrepresented relative to their abundance in live populations, making young and old individuals seem overly common. In species with low birth rates, such as whales, the age profile may either be U-shaped or L-shaped, depending on whether mortality rates increase significantly with age. Such profiles are termed "attritional" because they are the products of starvation, accidents, most types of disease, or other factors which preferentially affect very young and very old individuals in the population. In the context of prehistoric hunter-gatherers, such assemblages could result from scavenging dead animals or from hunting of the most vulnerable (young and old) individuals.

A number of factors may seriously compound the potential variation in mortality profiles, particularly in long-lived mammalian species (e.g., Haynes 1985; see also Chapter 1). The theoretical profile models for the age structure of live populations, and nonselective (catastrophic) and attritional mortality, nevertheless provide a useful point of departure for examining Thule Eskimo-bowhead whale relationships. Expectations regarding mortality profiles representing scavenging of naturally stranded animals versus those representing active hunting from a live population are discussed in more detail below.

Naturally Stranded Whale Assemblages

Of the various causes of mortality among bowhead whales, only one, ice-entrapment, would potentially result in catastrophic mortality profiles. (Mass 'live' beach strandings, while common among some whale species, has not been reported for bowheads.) Due to the bowhead's proclivity for remaining near ice margins and within broken ice fields, ice-entrapment can be expected, resulting in drowning or fatal crushing. Although there is little known evidence for significant mortality directly related to this cause (Mitchell and Reeves 1982), the absence of such information may be more a function of the early historic decimation of bowheads. Accordingly, we do not know the extent to which catastrophic ice-entrapment may modify broader patterns of mortality for the population as a whole. However, stranded bowhead profiles can be expected to show the entire range of bowhead whale age/size classes, with higher proportions of 'prime' adults represented when ice-entrapment is significant. Accordingly, if Thule Eskimos were scavenging such populations, mortality profiles from animals represented at Thule sites should resemble those of naturally stranded populations. Information in this instance is required from both processing and residential sites, since the argument could be made that any "selectivity" revealed at residential sites may simply be reflecting Eskimos' preference for particular bone sizes used in dwelling construction.

Active Hunting

In the case of active hunting, ethnographic accounts of historic bowhead whaling by North Alaskan Eskimos — the direct descendants in northern Alaska of Thule Eskimos — document a definite bias toward younger individuals on the part of hunters. Younger individuals are the easiest to manage during both pursuit and processing (see McCartney and Savelle 1985 and references therein). Similar reasons have been suggested for the apparent selection of smaller size individuals of other megafauna such as mammoths (e.g., Haury *et al.* 1959). Since Canadian Arctic Thule Eskimos exhibited similar technological, logistical, and social characteristics as North Alaskan Eskimos (see McCartney 1980b, 1984; McCartney and Savelle 1985; Savelle and McCartney 1990), Canadian Thule bowhead whale assemblages may be expected to result in mortality profiles that exhibit a disproportionately greater number of individuals in the younger/smaller age/size categories than that predicted by the theoretical living-structure model.

MATERIALS AND METHODS

Canadian Arctic bowhead whale bone assemblages from two different contexts are examined in this study: (1) a series of whales that were naturally stranded on Holocene raised beaches that pre-dated Thule Eskimo presence; and, (2) a series of whales found at Thule sites.

Naturally Stranded Bowhead Whale Assemblage

Studies of naturally stranded bowhead whales in the Canadian Arctic have been used primarily to determine Holocene sea ice conditions and/or isostatic rebound rates, which in turn have been used to generate models of Holocene paleoenvironments (e.g., Barr 1971; Blake 1975; Dyke 1979, 1980; Dyke and Morris 1990). Because these studies have focused on questions about the absolute abundance and distribution of bowhead whale skeletal elements, mortality profiles have not been determined. However, Dyke (personal communication 1989) did take single measurements on crania from 129 Holocene bowhead whales on Brodeur Peninsula, Baffin Island, immediately east of the area in which the Thule sites are located (see Fig. 10.2). The stranded whale remains are radiocarbon dated to between 10,500 and 1,000 B.P., and are thought to have been deposited as floating carcasses at or close to the respective contemporaneous shorelines. Subsequent isostatic rebound has elevated these bones to their present positions on raised beaches. These whales are presumably derived from the same Davis Strait stock from which later Thule Eskimos acquired whales. Dyke's sample presently is the only one available for comparison of mortality profiles with Thule material.

Archaeological Samples

Archaeological whale bones occur in many Thule sites in the eastern Canadian Arctic, either as structural components of semisubterranean dwellings at residential sites (see above) or as primary subsistence refuse at initial processing and caching localities. McCartney (1978) and E. Mitchell initially identified and measured bowhead bones at Thule sites in two regions on southeastern Somerset Island (regions 1 and 2, Fig. 10.2), while Savelle (1989) and McCartney identified and measured additional bones in region 1 and in nine adjacent regions (regions 1 and 3-11, Fig. 10.2).

It is assumed that all bowhead bone elements at these sites were derived from individuals of the Davis Strait stock, for which the study region represents the extreme western summer feeding range. Based on annual sea ice patterns (Lindsay 1975, 1977, 1981), the recorded historic

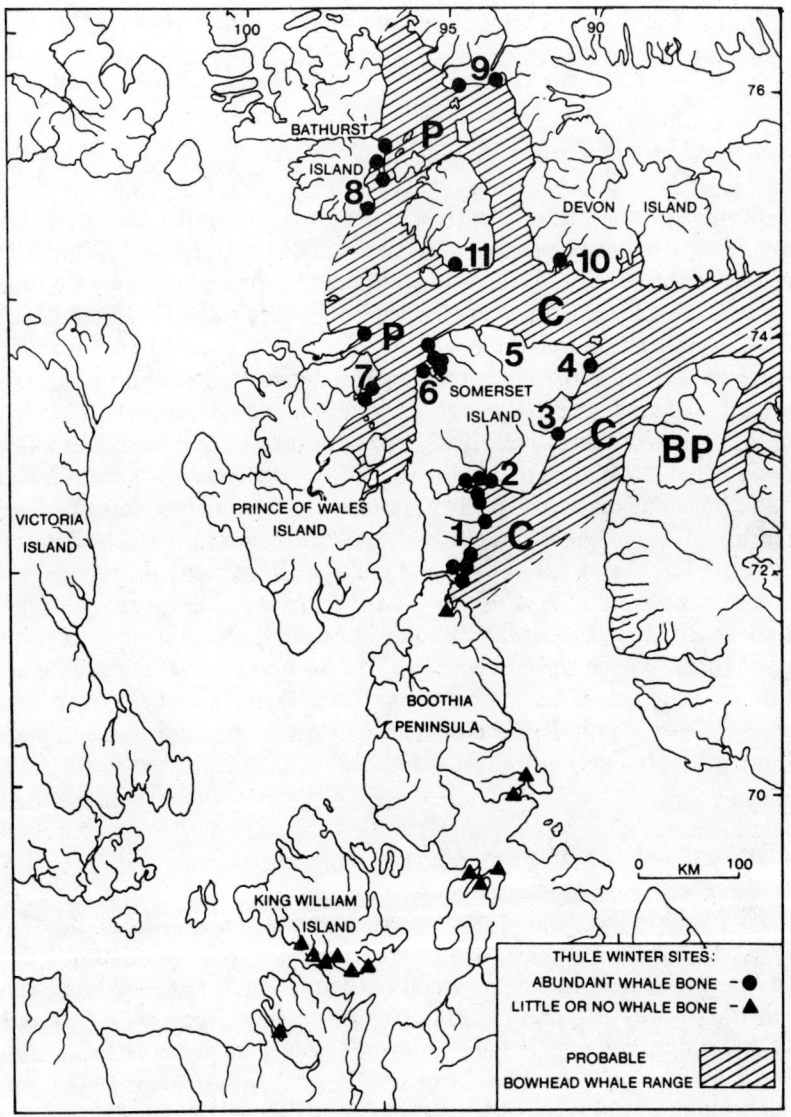

Figure 10.2. Thule Eskimo site distribution and archaeological bowhead whale bone abundance relative to bowhead whale "core" (C) and "peripheral" (P) summer ranges. Here, B.P. refers to the Brodeur Peninsula, where stranded Holocene bowheads occur. Numbers indicate general survey regions.

distribution of bowhead kills (Reeves *et al.* 1983), and recent sightings (Mansfield 1971; Reeves *et al.* 1983), the study region actually may be subdivided into *core* and *peripheral* summer ranges (Fig. 10.2). Core ranges are those in which bowhead abundance would be highest and most predictable, since sea ice clears earlier and more completely. Peripheral ranges are those which bowheads would frequent to a lesser degree and on a less predictable basis, due to later and often incomplete ice clearing. Bowheads appear to favor sea ice conditions of up to 50% loose pack (Brueggeman 1982).

A total of 10,477 bowhead whale elements were recorded at Thule sites throughout the study areas during the 1978 and 1988 field seasons. Using an individual site aggregation method, a minimum of 1234 individual bowheads are represented (cf. Grayson 1973). Using a total aggregation method for each entire survey region, on the other hand, a minimum of 963 individual bowheads are represented. From the total sample, 443 crania, 760 mandibles, 219 scapulae, and 122 cervical vertebrae were measured for size/age determinations (see below).

Determination of Bowhead Whale Size and Age from Skeletal Measurements

Traditional methods for constructing mammalian age profiles, such as tooth eruption-wear and dental annuli, are inappropriate for bowheads because they do not possess teeth, instead feeding through a series of baleen plates. Thus, a series of size measurements of selected bone elements was devised to estimate total whale length and, on this basis, the degree of maturation/age.

The measurement series was designed by E. Mitchell (Arctic Biological Station, St. Anne-de-Bellevue, Quebec), based on 16 cranial, 12 mandibular, 5 scapular, and 5 cervical vertebral measurements (McCartney 1978). In order to relate the size of these disarticulated elements to total animal length, McCartney measured a series of 14 Alaskan bowhead whale skeletons housed at the Los Angles County Museum. This control collection was assembled by Dr. Floyd Durham of that institution and represents the largest known for any bowhead population. Measurements of the Durham collection of skeletons were used to generate multiple linear regressions for animal length (McCartney 1980a; McCartney and Mitchell 1988). These regression models were then applied to the measurements from archaeological whale elements to estimate whale length. For the naturally stranded bowhead whale elements examined by Dyke, only maximum cranial width measurements are available. These were the only bones not directly measured by us, yet Dyke's descriptions of his measurements are consistent with our own, justifying application of the regression models for cranial width.

To determine age from whale length, measurements on recently killed Alaskan bowheads were employed (e.g., Maher and Wilimovsky 1963; Marquette 1976; Nehrini *et al.* 1984; Breiwick *et al.* 1984). Based on these measurements, the following size/age categories of live length are used: calves=4.0-6.0 m; yearlings=6.0-9.4 m; subadults (2-5 years)=9.4-12.0 m; adults=12.0+ m (males) and 14.0+ m (females).

A potential problem in extrapolating animal age from animal lengths, as opposed to more direct aging methods such as incremental structures and tooth eruption and wear, is that length profiles cannot be linearly correlated with age. That is, growth rates are highest immediately following birth and decrease throughout life. This has the effect of "damping" younger age category frequencies and artificially enhancing older age category frequencies relative to a theoretical living-structure model and thereby also has consequences for comparisons to catastrophic and attritional mortality profile models. This effect can be controlled for individuals up to and including the subadult class, since both lengths and ages are known. It cannot, however, be directly controlled for adults, since there is no independent method for assessing the ages of these individuals (although Mitchell [1984] reports probable growth layers in the periosteal bone of the auditory bullae). We simply do not know the full potential lifespan of bowhead whales, so that interpretation of mortality profiles on other than original whale size and, secondarily, major age categories is suspect.

RESULTS

The results of the size determinations for bowheads represented by the stranded and archaeological bone populations, and of a recent investigation of body length frequencies in a live bowhead population, are summarized in Figs. 10.3 and 10.4. Each of these is discussed separately below.

Live Bowhead Population

There are no available data on whale length frequencies in live populations of the Davis Strait bowhead stock. In the western Arctic however, a

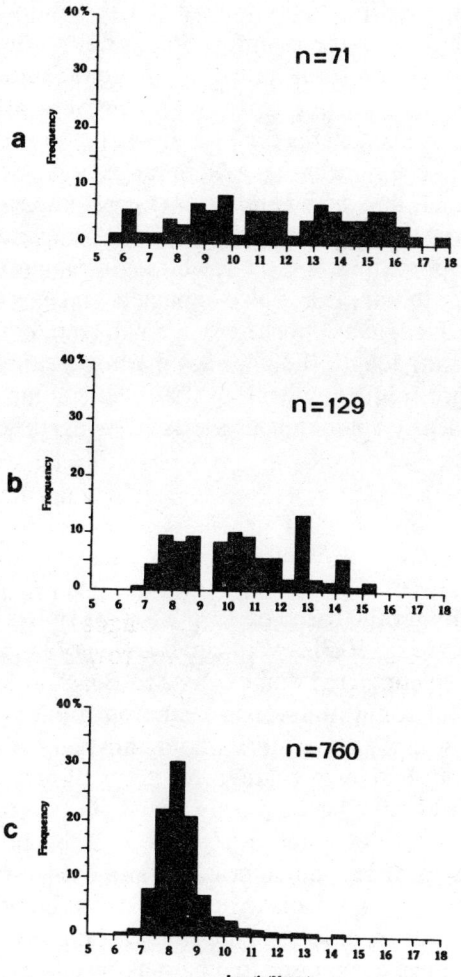

Figure 10.3. Relative length frequencies for bowhead whales in (a) live populations in four adjacent regions of the Beaufort Sea (after Cubbage and Calambokidis 1987:Fig. 2), (b) stranded Holocene population (ca. 10,500-1,000 B.P.), Brodeur Peninsula, Baffin Island, and (c) Thule sites in all survey regions indicated in Fig. 10.2.

recent study by Cubbage and Calambokidis (1987) provide length frequency data for the Beaufort Sea bowhead populations. This study employed aerial stereophotogrammetry to assess individual whale size in four adjacent regions in the Beaufort Sea. They recorded over 200 individual whale images on film from 7 August to 6 September, 1983. After correcting for measurement quality and potential whale duplication, they were able to determine lengths for 71 individuals (Fig. 10.3a).

It is immediately evident that a wide range of whale lengths are represented, and further, that there is no distinct peak among the smallest whale sizes (i.e., calves and yearlings: 4.0-9.4 m). The wide range of sizes is expected, and approximates the known length range for bowheads. The lack of a distinct initial peak is also expected, since, as noted earlier, profiles of bowhead length as opposed to age will tend to 'dampen' this peak. If calf and yearling length (i.e., 4.0-9.4 m) frequencies are combined, an initial age/length frequency peak of 31% results, and the modified age/length profile clearly approximates a classic living-structure (catastrophic) mortality model.

Stranded Bowhead Whale Sample

The mortality profile based on length for the stranded bowhead assemblage is shown in Fig. 10.3b. Very small and very large whale lengths are underrepresented compared to the censused Beaufort Sea live population. This *may* reflect some imprecision inherent in the regression models employed and/or in the aerial stereophotogrammetry. Otherwise, yearling, subadult, and adult size/age categories are well represented. Although there is no initial peak based purely on whale length, combining calf-yearling lengths results in an initial peak of 33%, and the modified age/length mortality profile again approximates a living-structure model. On the basis of these comparisons, we suggest that bowhead mortality and subsequent strandings, at least in the study area, probably result primarily from ice-entrapment as opposed to other natural causes.

Archaeological Sample

For the purposes of this discussion, we focus specifically on size determinations based on mandibles, since these offer the highest numbers for comparison between core and peripheral ranges. Examining first the overall mortality profile of bowheads represented at the Thule sites (Fig.

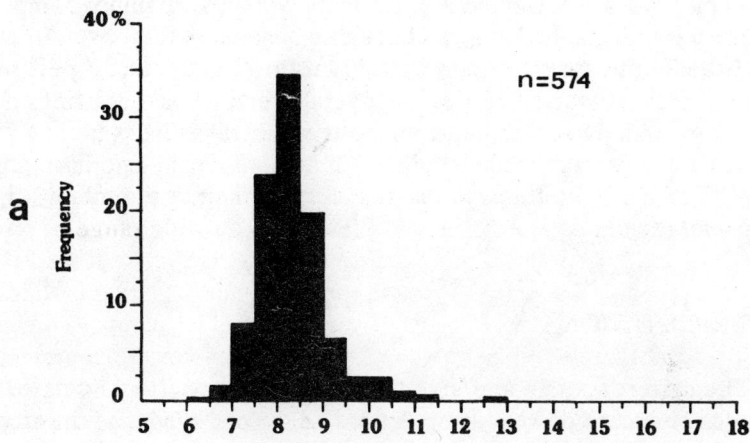

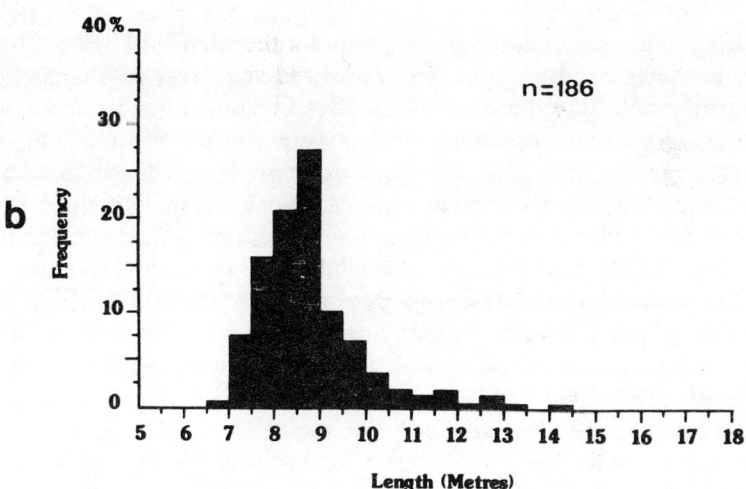

Figure 10.4. Relative length frequencies of bowhead whales represented at Thule sites in (a) core summer range, and (b) peripheral summer range.

10.3c), it is immediately evident that the observed pattern resembles neither the live nor the stranded whale profiles. Instead, the vast majority (91%) of whales represented are yearlings, with a very minor component (9%) representing subadult and adult size categories.

If the Thule sample is broken down into the core and peripheral regions (Figs. 10.4a and 10.4b), the overall trend toward yearlings is still evident in both. However, there are more subtle differences in the relative frequencies of subadults and adults. While these size categories represent only 7% of the individuals in the core region, they represent 18% in the peripheral region.

INTERPRETATIONS

The differences in age/size frequencies in the live and naturally-stranded Holocene bowhead populations on the one hand, and the archaeologically derived populations on the other, are striking. As discussed above, the naturally-stranded assemblage is typical of catastrophic episodes, probably resulting from ice-entrapment, while long-term attritional mortality (although possibly represented) does not appear to have been as significant for beached bowhead death populations specifically. If Thule Eskimos were scavenging stranded whales, we would expect living-structure (catastrophic) mortality profiles to be represented. Instead, the culturally-associated mortality profiles in Thule sites indicate a very definite selection of yearlings and, in the peripheral regions, also some subadults and adults. This overall selectivity in favor of smaller whales is not unexpected, given the ethnographically-documented selection of such animals by bowhead whale-hunting North Alaskan Eskimos. Further, this selectivity relates to live prey manageability from the human perspective.

The somewhat higher proportions of subadults and adults at Thule sites in the peripheral region may be significant. The overall whaling success rate, as measured by the ratio of total numbers of bowheads to total estimated Thule populations in these areas (Savelle 1990), is considerably lower than in the bowhead core region. It would appear that there was a tendency for Eskimo hunters to be less selective in these peripheral regions where bowhead numbers were not as abundant nor their seasonal occurrence as predictable. While the differences are subtle relative to the contrasts noted between culturally-associated and naturally stranded mortality patterns, exclusive selection of small whales may have been most practical only within the most productive whaling areas, where prey were concentrated.

CONCLUSION

The data presented here strongly suggest active, selective hunting of young bowhead whales by Thule Eskimos. These Eskimos took primarily yearling animals, secondarily subadults, and only rarely adult animals. These data also demonstrate regional variability in prey selection according to size. This in turn may relate to regional variation in bowhead whale abundance and predictability.

In some respects the Thule-bowhead whale predatory relationship is unique in that it relates to (a) the extremely large size of this prey species, (b) harvesting in a marine rather than terrestrial environment, and (c) the use of bone elements in an architectural context in addition to procurement of meat, blubber and entrails. Nevertheless, we feel the study may be instructive at a more general level for investigations of human-megafauna relationships in other contexts. These include, for example, Old World Upper Paleolithic sites, in which mammoth bones were used as structural components in shelters but their dietary importance is equivocal (e.g., Klein 1973; Soffer 1985), and New World Paleoindian sites, in which the nature of mammoth or mastodon procurement techniques (i.e., hunting vs. scavenging) remains a subject of considerable debate (compare, for example, Saunders 1980 and Haynes 1985).

ACKNOWLEDGMENTS

Logistical support for the 1978 and 1988 field investigations was provided by the Polar Continental Shelf Project (Department of Energy, Mines and Resources, Canada). Financial support for the 1978 investigations was provided by the Archaeological Survey of Canada (National Museum of Man) and the Department of Indian and Northern Affairs (Canada), and for the 1988 investigations by the Social Sciences and Humanities Research Council of Canada. To these organizations we extend our gratitude. Capable field assistance in 1978 was provided by Bryan Kemper, Allen Clarke, Michelle MacLaughlin, Edward Mitchell, and David Sudlovenick, and in 1988 by Karen Digby-Savelle, Elisa Hart and Terry Manik. Data entry and analysis of the 1988 measurements were competently carried out by Brian Schnarch. Edward Mitchell (Arctic Biological Station, Quebec) has been of great assistance over the years in providing cetacean expertise to our archaeological projects, and the late Floyd Durham generously provided facilities and assisted in McCartney's examination and measurement of the bowhead materials at the Los

Angeles County Museum. Finally, we wish to thank Arthur S. Dyke for supplying the measurements for the stranded Holocene bowhead elements and for his interest in our research.

11

Seasonality Studies and Paleoindian Subsistence Strategies

Lawrence C. Todd

INTRODUCTION

In the early 1970s, information on the seasonality of late Pleistocene/early Holocene bison bonebeds from the North American Plains began to be developed from studies of patterns of dental eruption and wear. By the 1980s, it became evident that the strategies involving seasonal use of bison by Paleoindian groups was not the same as that documented for later hunter-gatherer groups living on the shortgrass plains (Todd 1987a). In particular, Frison (1982d:194) has observed that ethnographic information and archaeological data from the Late Prehistoric Period indicate many of the large kills took place in the fall and involved a "strategy of processing bison for winter use by drying, pemmican manufacture, or some related techniques." Evidence for processing areas typically found associated with Late Prehistoric bison kill sites includes several types of

> readily identifiable features and artifacts. These include stone-heating pits, stone boiling pits, and piles of bone reduced to varying sizes for boiling out of the bone grease. Anvil stones and hammerstones were used extensively in bone crushing and breaking processes.... During the early fall, the bison are in the prime condition that provides the thick layer of back fat used in pemmican manufacture. Both dried meat and pemmican provided the necessary surpluses to insure winter survival (Frison 1982d:200).

Paleoindian sites are markedly different from later sites. In contrast to the well-documented pattern of kills and associated processing features in Late Prehistoric contexts, Frison notes that "attempts to apply the same model to Paleo-Indian bison kills have not been successful" (1982d:200). Specifically, none of the known Paleoindian sites contain processing features like those described from later sites such as Ruby (Frison 1971), Piney Creek (Frison 1967), Big Goose Creek (Frison *et al.* 1978), Head-Smashed-In (Brink and Dawe 1989; Brink *et al.* 1986), or Bugas-Holding (Rapson 1990; Rapson and Todd 1989; Todd and Rapson 1988b). Hearths at Paleoindian sites instead tend to be small ephemeral features, and the associated bone assemblages do not indicate intensive grease or marrow processing activities (Jodry 1987:143-144).

Rather than representing early fall mortality events, the seasons in which Paleoindian bonebeds were formed tend to cluster in late fall through early winter. This is a time of year when the modern bison's condition, particularly fat reserves, are quite poor (Frison 1982d:201; Speth 1983, 1987; Speth and Spielmann 1983), but it is also a time of the year when preservation of meat by freezing rather than drying would have been possible, and when mobile prey can provide a predictable key resource. Considered together, these sources of evidence along with the distributional patterns within several bonebeds led Frison (1982d) to suggest that Paleoindian winter subsistence may have involved the use of frozen meat caches.

The value of concentrated, predictable food sources for winter survival on the recent Plains cannot be denied. However, one of the problems with the frozen meat cache model is that "one element would have been lacking, which is the fat of the prime animals . . . killed in the early fall of the year" (Frison 1982d:201). The nutritional importance of dietary fats and carbohydrates as supplements to lean meats has been highlighted on a number of occasions by Speth (Speth 1983, 1987; Speth and Spielmann 1983), yet new evidence for seasonal use of kill sites by Paleoindians continues to point primarily to late fall-winter mortality. Although the number of sites representing other seasons of the year has increased (Todd *et al.* 1990), the overriding tendency for Paleoindian bonebeds to represent what might be considered a "fat-indifferent" winter use of animals still stands.

Several alternatives are considered below in an attempt to understand this unexpected pattern of Paleoindian seasonal resource utilization. The goals of this paper are twofold. I first outline some of the current evidence on seasonal patterns of Paleoindian bison use. I then raise some questions about the relationships between late Pleistocene/early Holocene resource structure and possible human adaptive responses to it. As highlighted by Gamble's (1987) engaging discussion of Middle Pleistocene colonization of Europe, the most appropriate units of analysis for Paleoindian studies may well be "regions and the variable structure and distribution of resources that they contained and that patterned human behavior" (1987:95) rather than sites or artifact classes.

BISON DENTITIONS AND SEASONS OF DEATH

Estimations of the seasons of death in prey here are inferred from tooth eruption and wear. Modern comparative specimens, coupled with measurements of molar crown heights on archaeological specimens, has become a common approach for investigating Plains bison bonebeds (Frison 1978a, 1982a; Frison and Reher 1970; Frison et al. 1976; Frison et al. 1978; Reher 1970, 1973, 1974; Reher and Frison 1980; Todd 1987b; Todd and Hofman 1987; Todd et al. 1990; Wilson 1974, 1980, 1983; Wilson et al. 1988). Table 11.1 lists Paleoindian bonebeds for which seasonality data are available. The geographic locations of many of the sites are shown in Fig. 11.1.

Multi-animal bonebeds yield age data for relatively large numbers of individuals in each age group. These kinds of samples often permit relatively secure assessments of the seasons in which animals died, as well as considerable assurance that "missing" seasons in fact reflect the temporal boundaries in which deaths occurred rather than the effects of sampling error. The techniques of estimating dental ages are based on a combination of eruption and attrition stages, which prove more suitable than reliance on either eruption or attrition alone.

Estimated seasons of death based on mandibular dentitions are most commonly reported in 0.1 year increments, beginning with an assumed spring birth pulse (Table 11.1). These data are summarized by site assemblage in Fig. 11.2. Using these data as a basis for examining the range of seasonal mortality, a frequency of occurrence per 0.1 year segment was tabulated by counting the number of assemblages spanning each increment. If seasonality data for a particular site covers a multi-increment span, it is counted once in each. For example, the Agate Basin 2 assemblage, which ranges from N+0.6 to N+0.9 yr, is counted once in the N+0.6-N+0.7 increment, once in the N+0.7-N+0.8 increment, and once in the N+0.8-N+0.9 increment. Cases involving shorter durations that span a boundary between increments, such as Perry Ranch, Casper, and Finley, are counted in both.

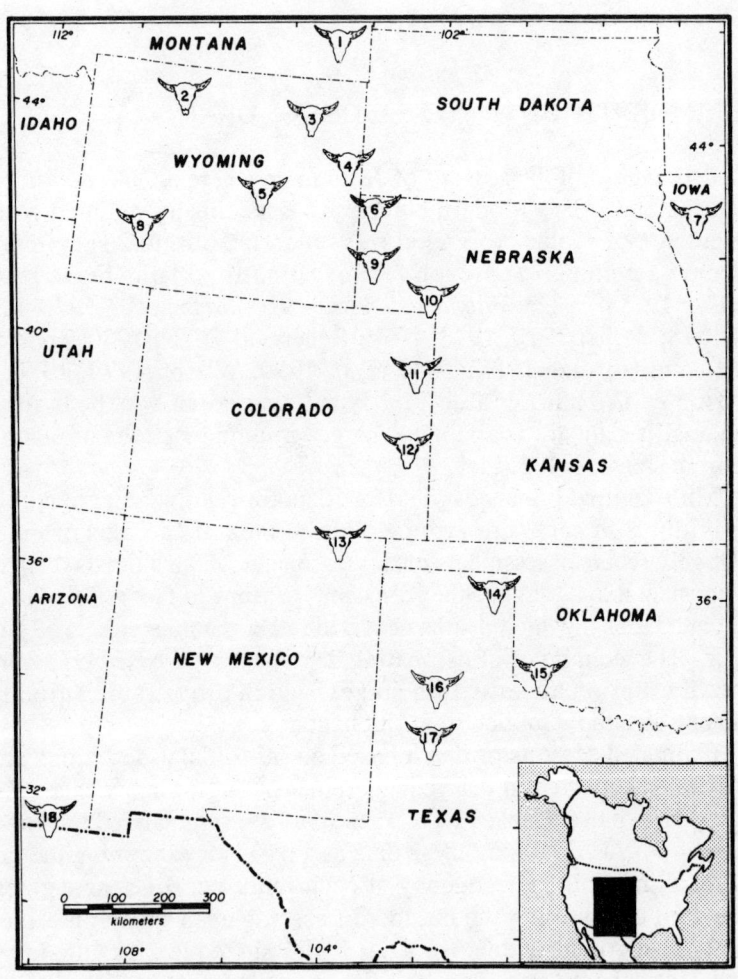

Figure 11.1. Locations of selected Paleoindian bonebeds for which seasons of death have been established. Numbers refer to cases listed in Table 11.1.

Table 11.1. Dental ages and estimated season of death at Paleoindian bison bonebeds.

#	site	dental ages[a]	season[b]	source
4	Sheaman[c]	N+0.1 to N+0.25 yr	late spring-early summer	Frison 1982a:260
1	Mill Iron	N+0.9 to N+1.1 yr	spring to early summer	Todd and Rapson 1988a
9	Scottsbluff	N+0.1 to N+0.3 yr	late spring-summer	Todd et al. 1990
10	Clary Ranch	N+0.3 yr	late summer	Myers et al. 1981
16	Plainview	N.0.3 yr	late summer	Fawcett 1987:402
12	Olsen-Chubbuck	N+0.3 to N+0.5 yr	late summer or early fall	Frison 1978a; Wilson 1974
14	Lipscomb	N+0.3 to N+0.5 yr	late summer or early fall	Todd et al. 1990
6	Hudson-Meng	N+0.5 yr	fall	Agenbroad 1978
15	Perry Ranch	N+0.6 yr	late fall to early winter	Saunders and Penman 1979
5	Casper	N+0.6 yr	late fall to early winter	Reher 1974; Wilson 1974
8	Finley	N+0.6 yr	late fall to early winter	Todd and Hofman 1987
13	Folsom		late fall or early winter	Frison 1978b:113
	Frasca		late fall to early winter	Fulgham and Stanford 1982:5
2	Horner: Horner I	N+0.6 yr	late fall to early winter	Todd and Hofman 1987
	Horner II	N+0.6 yr	late fall to early winter	Todd 1987b
17	Lubbock Lake: FA6-11	N+0.6 yr	late fall to early winter	Johnson and Holliday 1980
	FA6-3[d]	N+0.8 to N+0.9 yr	spring	Johnson and Holliday 1981
	Jones-Miller		winter	Stanford 1979
	Rex Rodgers		winter	Speer 1978:119
11	Mona Lisa: Locality A	N+0.6 to N+0.7 yr	early winter	Wilson 1983:352

(continued)

Table 11.1. (Continued)

#	site	dental ages[a]	season[b]	source
3	Carter/Kerr-McGee: Cody-Alberta	N+0.6 to N+0.7 yr	early winter	Frison 1984
18	Murray Springs[e]	N+0.7 to N+0.8 yr	winter	Wilson et al. 1989; Frison 1978a
7	Cherokee: Horizon IIIa	—	mid- to late winter	Pyle 1980:183
4	Agate Basin (Area 1): Agate Basin	N+0.5 to N+0.9 yr	winter to early spring	Frison 1982a
4	Agate Basin (Area 2): Hell Gap	N+0.6 to N+0.7 yr	winter	Frison 1982a
	Agate Basin	N+0.6 to N+0.9 yr	winter to early spring	Frison 1982a
	Folsom[f]	N+0.9	late winter/early spring	Frison 1982a
12	Mile Creek	—	late winter/early spring	Rodgers and Martin 1984

- Label in Fig. 11.1.

[a] The range of dental ages represented in all younger age groups.
[b] Season based on the most common dental age rather than the age range.
[c] Only three teeth are represented at Sheaman, making seasonality determination indefinite.
[d] Seasonality estimate is based on three fetuses at "3/4 to near term" (Johnson and Holliday 1981:185).
[e] The sample from Area 4 (the Multiple Bison Kill) is small but displays tightly clustered age groups.
[f] A single calf and a small amount of fetal material suggests the season of death, but data are not conclusive.

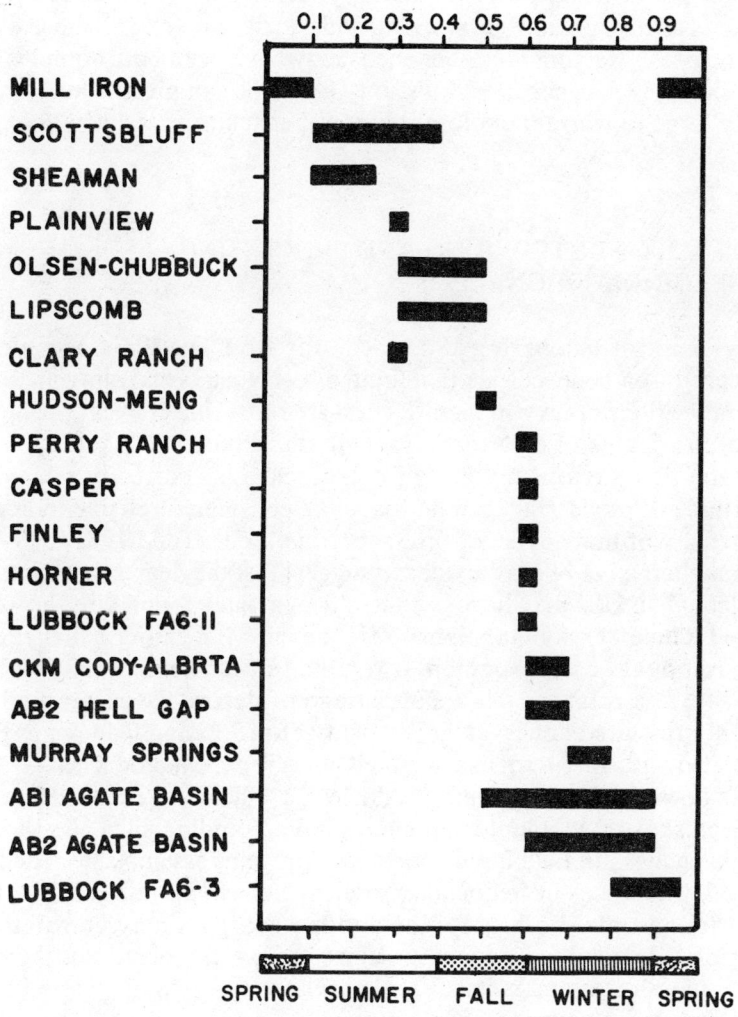

Figure 11.2. Seasons of death based on eruption and wear patterns of mandibular dentitions from Paleoindian bonebeds.

Tabulated in this way, a total of 39 seasonal increments are represented overall. These data are then converted to percentage by increment values and plotted in a seasonal "rose" diagram (Fig. 11.3). This figure clearly illustrates the predominance of late fall and early winter death assemblages for the late Pleistocene/early Holocene. While the possibility of error is always great when working with small samples, and the geological processes affecting these bonebeds may have been both complex and unpredictable (Albanese 1978; Frison 1984), the sample at hand is sufficiently large to warrant exploration now, particularly as a way to refine current hypotheses.

EVIDENCE OF BUTCHERING AND PROCESSING IN PALEOINDIAN BONEBEDS

Evidence of butchering and processing from late Pleistocene/early Holocene bison bonebeds is often limited, as cutmarks and other unambiguous indications of human modification are rare. The conventional equation of the degree of disarticulation with the amount of butchery, or the assertion that the total amount of bone breakage is directly related the magnitude of processing, can no longer be considered reliable indicators of intensity of human use of products from a kill (Todd 1987a, 1987b). Several alternative lines of evidence suggest that the degree of processing at Paleoindian kills may have been less than at later period kills, however. These include (1) skeletal element frequencies indicating either limited bone removal or non-segmental removal of entire limb units (Todd 1987a), (2) a relatively low percentage of definitely cutmarked and humanly fractured bones, and (3) paucity of processing features or those associated with long-term use of stored or cached meat products.

As documented elsewhere (Todd 1987a), there are clear differences between skeletal element frequencies at Paleoindian sites and those at most later sites. In Late Prehistoric northern Plains assemblages, one finds marked differences in frequencies of elements *within* limbs. In the front leg, for example, both scapulae and metacarpals may be relatively common, while humeri and radius-ulnae are less frequent. For the hindlimb, innominates and metatarsals are relatively common, while femora and tibiae are infrequent.

At Paleoindian sites, all bones of each limb tend to be represented in approximately equal proportions; removal of bones from the sites seems to have been as complete limb units rather than as segmented subsets. This is not the sort of segmental butchery that is normally associated with

Seasonality and Paleoindian Subsistence

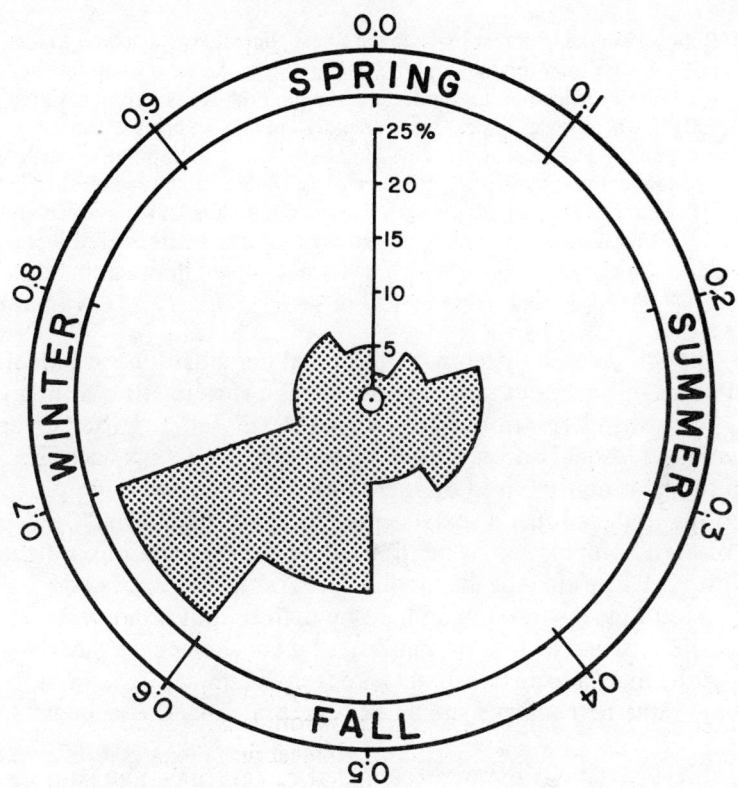

Figure 11.3. Seasons of death in Paleoindian bison assemblages, based on the frequency of occurrence across 0.1 year increments.

immediate selective bulk-processing for stores, and therefore contrasts with processing strategies evident at Late Prehistoric kills, where large quantities of marrow and bone grease were rendered.

Speth (1983, 1987; Speth and Spielmann 1983) and others (e.g., Spiess 1979:26-29) observe that seasonal variation in fat content is related, in part, to the sex of the game animal. Seasonal changes in bison condition are summarized as follows:

Both sexes are in poorest shape in the spring, but males reach their lowest point earlier than females and are improving at the time when females reach their lowest point. Similarly, although both sexes improve during early summer, males improve more rapidly. In mid- to late summer, however, males decline sharply while females continue to improve gradually throughout the summer, fall, and even early winter. Both sexes decline in late winter, but males often enter this period with less fat reserve and are therefore more vulnerable to undernutrition if conditions are severe. Males, however, rebound much faster and sooner in the spring than pregnant or lactating females (Speth 1983:163).

Although data on differences in sexual composition for Paleoindian kill assemblages are not available for enough sites to allow a full evaluation, present information suggests that Paleoindian groups were not unaware of seasonal differences in carcass quality. Estimates of sex ratios are most commonly based on metapodials. If no selective processing according to the animal's sex was practiced, we should find that, regardless of herd composition at the time of death, the sex ratio indicated by forelimbs matches that of the hindlimbs. This assumes, of course, that no significant biases were introduced by differential bone destruction or deliberate removal of fore or hindlimbs by consumers. In other words, if 30% of the metacarpals (forelimbs) are from males and 70% from females, then the same percentages should be represented by the metatarsals (rear limbs).

Table 11.2 shows the percentages of metacarpals judged to be from males, minus the percentage of male metatarsals. This provides a rough measure of the "sexual equability" of the lower front and hind legs (Fig. 11.4). At the Agate Basin site, the percentage of mature male metacarpals is much higher than for the metatarsals. Given the apparent non-segmental butchering pattern for limbs noted earlier, coupled with the naturally higher food yield (fat and protein) of the rearlimb in relation to the forelimb (Binford 1978; Metcalfe and Jones 1988), selective deletion of male rearlimbs is suggested at this site.

In contrast, the percentage of male metacarpals is much lower than the percentage of male metatarsals at Olsen-Chubbuck, Scottsbluff, and, to a lesser degree, Casper. In other words, there are *more* male rearlimbs than would be expected at these sites, probably due to selective removal of female rearlimbs. The samples available for comparison are very small and by no means provide unambiguous indicators of selectivity, yet differential frequencies suggest selective removal of male rearlimbs from the Agate Basin site (associated with winter to early spring mortality) while the two summer-to-early fall sites (Olsen-Chubbuck and Scottsbluff)

Table 11.2. Relative frequencies of mature male metapodials from several Paleoindian bonebeds.

site	metacarpals		metatarsals		
	N	% male	N	% male	% difference
Agate Basin	16	50	6	25	25
Casper	6	11	7	18	-7
Finley	35	73	23	72	1
Lipscomb	12	22	9	23	1
Olsen-Chubbuck	10	30	12	48	-18
Scottsbluff	3	14	7	27	-13

Sources: Agate Basin (Zeimens 1982:Fig. 4.9); Casper, Finley, and Olsen-Chubbuck (Bedord 1974:Table 6.19); Lipscomb and Scottsbluff (Hillerud 1970:Table 5).

suggest selective removal of female rearlimbs. This pattern would be expected if removal of carcass segments were, to some degree, a function of animal nutritional state. While extensive processing for fat and bone grease may not have taken place during the Paleoindian period, the metapodial data suggest that differences in body fat apparently did play at least some role in butchery/transport decisions.

The prevalence of late fall through early winter mortality at most Paleoindian bison bonebeds makes Frison's (1982d) suggestion that some of the carcasses or carcass segments were frozen and cached seem quite reasonable. Freezing would certainly increase the potential use life of the meat products. But are there indications that Paleoindians made use of this potential? It is possible, after all, to strip meat from carcasses during a warm season kill, leaving few cut marks and large articulated units. Because cold weather causes carcasses to stiffen more quickly and the parts remain stiff for many months, the probability of direct evidence of butchery should increase in contexts where people are using frozen rather than fresh carcasses. Thus, there should be *more* evidence of butchery on bones rather than less, if parts are kept in frozen caches.

As noted in the introduction of this chapter, one of the more striking differences between Paleoindian bonebeds and those of later periods is the lack of associated processing areas or related specialized features. The multiplicity of hearth forms, stone-filled features and large quantities of

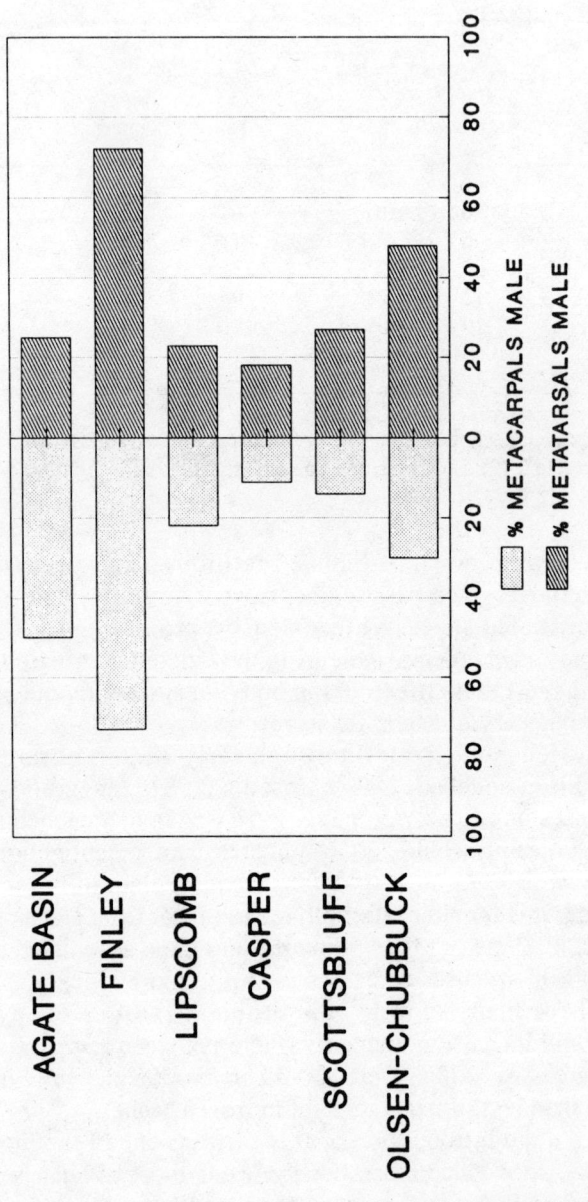

Figure 11.4. Differences in the percentages of mature male metacarpals and metatarsals relative to mature females.

fractured bones so common in processing sites of later periods are not known from Paleoindian sites. Instead, hearths are usually very shallow unprepared features, such as those reported by Frison from the Folsom (see Frison 1982b:Figs. 2.40, 2.43) and Hell Gap (Frison 1982c:Figs. 2.81, 2.84) levels at the Agate Basin site (see also Wilmsen and Roberts 1978: Fig. 166), or they are represented by very sparse scatters of charcoal. Evidence of burning in Paleoindian contexts therefore points to very short term surface fires, such as the "burned area" identified at the margins of the Horner II bonebed (Todd 1987b:Fig. 5.18) and those at Horner I (Todd et al. 1987:Fig. 3.6) and Hudson-Meng (Agenbroad 1978). Moreover, while we now have at least some bonebed sites representing other or all seasons of the year (Table 11.1), there seems to be no seasonal variation in the patterns of processing. It is possible that processing may not have taken place directly at the Paleoindian death sites, but at present there is little archaeological evidence to support this suggestion.

Returning to what we know about more recent Plains peoples, one of the primary goals of processing fall kill products during the Late Prehistoric period was to extract bone grease and marrow for making *pemmican*, a concentrated energy-rich food that can be set aside for later use. Dried meat/protein was also an important storable product from the kills, but the dietary significance of fat and grease in the over-wintering strategies of Late Prehistoric northern Plains Indians is often overlooked.

In his discussion of processing features at the Piney Creek sites, Frison (1967:35) reviews several ethnographic accounts of pemmican manufacture. Of particular note is Thompson's (1962:312) "pemmican recipe" in which 20 pounds of "hard fat", 20 pounds of "soft fat", and 50 pounds of dried meat were combined to form a 90 pound bag of pemmican. Clearly the fat component of pemmican offers the greatest caloric value, and fats may have been more critical than the stored lean meats in this food. A sufficient supply of fat lasting throughout the winter allows humans to make use of relatively fat-depleted kills later in the year. Thus, stored lean meat, whether dried or frozen, is only a partial answer to the problem of over-wintering.

Paleoindians may have shown some selectivity in their removal of carcass segments from kill sites according to prey sex and nutritional state, but there currently is no indication that they were preparing and storing large quantities of fats. This anomaly can be explored in other ways, particularly through considerations of habitat structure and seasonal cycles that may have been unique to the Upper Pleistocene/early Holocene.

LATE PLEISTOCENE/EARLY HOLOCENE SEASONALITY

Present difficulties in interpreting seasonal mortality patterns apparent at late Pleistocene/early Holocene sites may be due to misconceptions arising from direct use of modern grassland ecosystems as analogs for reconstructing past environments. An increasing number of studies (e.g., Chomko and Gilbert 1987; Graham 1987; Graham et al. 1987; Johnson 1986; Semken and Falk 1987; Walker 1987; Wendland et al. 1987), using a variety of data sets, are beginning to present a picture of the structure of the late Pleistocene environment and human adaptations on the North American Plains, and it differs markedly from the modern pattern (Martin and Martin 1984, 1986). Although the annual temperature was somewhat lower overall, seasonal extremes were not as great as they are now; summers may have been cooler, but winters probably were warmer. One of the most compelling lines of evidence for this difference is the widespread presence of "disharmonious faunas", animal communities containing species that are today allopatric (Graham 1987; Graham and Lundelius 1984; Guthrie 1984). This implies that the spatial structure of resource availability in Pleistocene environments differed as well. Guthrie (1984) suggests a structure comprising "plaids" of diverse resource patches rather than the zonal "stripes" that today differentiate environments in the higher latitudes.

Guthrie (1980, 1984) has also argued that this difference in "seasonality" may be one of the factors responsible for the larger body-size of late Pleistocene/early Holocene bison. A longer "somatic growth season" resulting from a more extended period over which dietary protein was available (Guthrie 1984:267-270) may have been a key factor in the maintenance of the larger forms of bison. Repeated warnings that Pleistocene biotic communities do not have modern analogs also apply to the nature of Pleistocene seasonality. Thus, in attempting to characterize human adaptations of late Pleistocene, we cannot assume that our current perceptions of hunter-gatherer seasonal foraging patterns are directly applicable.

DISCUSSION: SHARING THE FAT OF THE LAND

Most of the primary biomass in grassland settings occurs in the form of grasses, plants that are high in cellulose and not directly suitable as human food (Bamforth 1988; Foley 1982; Reher 1977). On the other hand, these grasslands can support a rich herbivore fauna that can make

Seasonality and Paleoindian Subsistence

effective use of the grasses. Foley (1982:398) notes that "while grass may form the ideal food base for specialized herbivores (hence the increase in large mammal biomass), they do not provide a suitable resource for hominids, who require concentrated, high quality plant foods." Human use of grassland habitats depends upon extracting usable food energy through the herbivores, which convert the grasses into more digestible forms, particularly (meat) proteins and fats. Two commons ways to accomplish this are hunting as "a predation strategy dependent upon wild, indigenous mammals" and (in more recent periods of human history) pastoralism as "a parasitic strategy dependent upon domestic, exogenous large mammals" (Foley 1982:400). Human use of mammals on the North American Plains was strictly predatory, with an emphasis on bison. However, hunting patterns seem to have changed through time, as has the nature of processing strategies and the seasonal patterns of harvesting. By assuming that seasonal cycles in bison nutrition were like those for more recent bison populations, Paleoindian bonebeds indicate seasons of death primarily when they would seem to have been in relatively poor physical condition.

Speth has drawn attention to the dietary problems associated with seasonal use of fat-depleted (lean) meat (Speth 1983, 1987; Speth and Spielmann 1983). Modern hunter-gatherers' responses to these difficulties (Speth 1987:17; Speth and Spielmann 1983:18-21) include:

1) selective, fat-sensitive procurement of prey and adjusting processing strategies in response to the degree of fat depletion in prey;
2) building up of human body-fat reserves prior to the period of food stress;
3) procurement and storage of plant foods that are rich in oils or carbohydrates.

Ethnographically documented Plains hunter-gatherers solved problems of seasonally fat-depleted game through some combination of the three options. Fall procurement of large quantities of meat, and more importantly fat (stored products such as pemmican), fat-sensitive seasonal hunting and processing, together with fall feasting all seem to have been important in Late Prehistoric Plains economies.

The options for exchange with other human groups and/or plant cultivation were either very limited or absent for Paleoindian groups. Evaluation of the possibility of stored wild plant foods is difficult, but it *is* possible to examine the related question as to why the use of animal products by Paleoindian groups appear to have differed so markedly from later Plains peoples. One of the major differences between Late Prehistoric subsistence practices and what we now can infer about Paleoindian seasonal use of bison is that Paleoindians did not seem to process products from fall or early winter kills in such a way as to provide large quantities of storable fats. Kelly and Todd (1988) have suggested that the lack of evidence for extensive fat rendering by Paleoindians reflects late Pleistocene/early Holocene resource structure and concomitant mobility strategies. The restrictions that food storage strategies can place on mobility may have outweighed any potential benefits. In light of this possibility, we nevertheless must address the question of how these groups coped with the possible seasonal fluctuations in fat/carbohydrate availability.

There are a number of potential explanations for Paleoindians' apparent lack of concern with the procurement and storage of large quantities of fats and bone grease. One is that, given current understanding of late Pleistocene/early Holocene seasonality on the Plains, and particularly Guthrie's model of longer periods of plant productivity per annum, seasonal fluctuations in bison body fat condition may not have been as extreme as they were by the middle and late Holocene. The large bison of the late Pleistocene/early Holocene (circa 12,000-8,000 B.P.) almost certainly faced periods of severe seasonal resource stress (Guthrie 1980; Walker 1986), but the periods may not have been as long-lasting or as acute as those experienced by large Plains herbivores in later periods.

Figure 11.5 presents a model of these differences, in which the amount of time that animals are in a "fat-depleted" state decreases proportionately as the somatic growth period increases, in accordance with Guthrie's (1984) more general model. Using 11.5b as a comparative baseline, herds hunted by Paleoindian would have been most severely fat-depleted (Fig. 11.5c) over a shorter period than herds utilized by later Plains groups (Fig. 11.5a). This difference in seasonal resource structure may well have permitted systems of use by humans that were very different from those employed by later, ethnographically documented bison hunters (Fig. 11.5b). Thus, the intensity of physiological stresses experienced by bison populations and ultimately by human predators in the Americas during the late Pleistocene/early Holocene may have been significantly less than the stresses experienced by Late Prehistoric groups inhabiting the same region. Storage of fats in large quantities may not have been necessary if even small amounts of fats or carbohydrates could be secured over a longer part of the annual cycle.

Another important consideration is the way in which winter mobility of human groups is related to seasonal temperature and precipitation extremes. During the Late Prehistoric Period on the Northern Plains, mid-

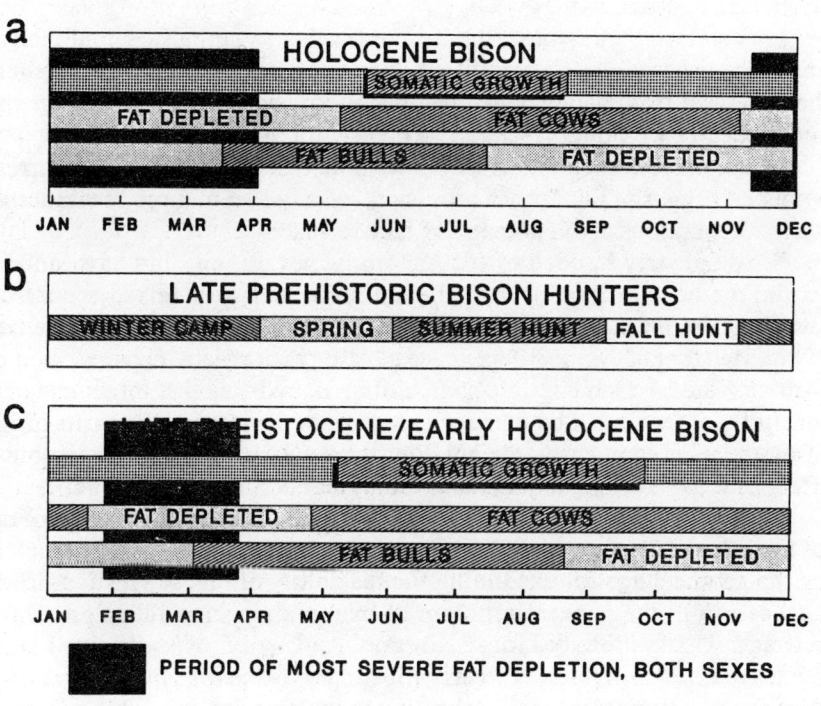

Figure 11.5. Model of relationships between somatic growth potential (Guthrie 1984:268) and length of time in which bison fat resources are available to human predators: (a) schematic representation of Holocene bison seasonal change; (b) seasonal variation in human use of bison as evidenced by the Blackfeet in the historic period (Ewers 1955:123-129); and, (c) proposed structure of Late Pleistocene/early Holocene bison seasonal variation in nutritional state.

winter mobility was often very limited due to deep snow, sub-freezing temperatures and high winds. During these periods, food procurement was nearly impossible, and some form of stored products was essential for survival. As with the intensity and duration of seasonal fat depletion in prey, periods of climatically-enforced low mobility that in turn limit resource procurement may also have been less severe during the late Pleistocene/early Holocene.

Frison (1982d:201) suggests that some other as yet unspecified dietary provision assured the availability of fats or carbohydrates to supplement the lean meat from winter-killed bison. Alternative sources of food energy automatically would represent a more viable option if the periods of bison fat depletion were shorter and winter mobility less restricted. If, as current evidence suggests, this "other provision" was not animal product storage, what other options could potentially take its place?

While clearly in need of further study, several authors have underscored the need to look beyond the notion of meat proteins as the single and constant dietary staple of Paleoindian groups (Kornfeld 1988; Meltzer 1988; Meltzer and Smith 1986), instead calling for greater consideration of plant resources. One possible source of carbohydrates that might enhance the utility of lean meat by humans comes from the stomach contents of the bison themselves. During the late Pleistocene/early Holocene, a combination of protein from lean meat and the byproducts of microbial fermentation from the rumen may have provided a viable alternative to fall storage of fats.

To set the stage for examining the feasibility of this suggestion, let us first return to the general problem of mammalian survival in grassland settings. Using primary biomass in grasslands requires specialization on the part of herbivores, as is clearly illustrated by the digestive anatomy of ruminants. Ruminants are no better at digesting the insoluble carbohydrates (cellulose and hemicellulose) in plant cell walls than humans, instead relying on micro-organisms harbored within the digestive system. Compartments therein facilitate a symbiotic relationship between certain micro-organism taxa and the ruminant host. Hungate summarizes this system in the following way:

> The abundance of carbohydrates in plant cell walls is the basis for the evolution of the cooperative model for animal-microbe relationships. The carbohydrate polymers of the wall, indigestible by most animals are digested and fermented by the microbial partner, with the waste fermentation products and the microbial bodies used by the host (Hungate 1984:2).

Seasonality and Paleoindian Subsistence

Dziuk describes the ruminant stomach as "a large, carefully controlled fermentation vat in which substrate (food and salivary constituents) and water are added and end products of metabolism from bacteria and protozoa are removed" (1984:331).

The modified forestomach of ruminants ensures that digesta remain in the stomach long enough for the fermentation process to take place and meanwhile protecting the micro-organisms working on the plant matter from the acidic digestive juices. In all, the ruminant stomach is divided into four compartments (Fig. 11.6). No digestive juices are secreted in the first three compartments (reticulum, rumen, and omasum), where "extensive fermentative digestion results from large numbers of bacteria and protozoa" (Dziuk 1984:320).

Two major sets of nutritive products are derived via the fermentation process. The first is a series of short-chain volatile fatty acids (VFA, principally acetate, propionate and butyrate) that are absorbed into the bloodstream and tissue, serving as energy sources and as the raw material for biosyntheses. The second nutritive product comes from the bodies of the micro-organisms themselves, which pass out of the ruminoreticulum and are digested to form additional amino acids, fatty acids and vitamins (Wolin 1981:Fig. 1). VFA are particularly important byproducts since acetate supplies 50-60% (Hungate 1966; cited in Orpin 1982:502) or even 70-80% (Eastwood and Brydon 1985:121) of the animal's energy requirements. The "propionate [which] is used for gluconeogenesis, accounting for some 50% of the glucose used by the ruminant" (Orpin 1982:502).

Humans also harbor micro-organisms in the right side of the colon that produce VFA through microbial fermentation of non-digestible plant materials (Eastwood and Brydon 1985:113; Wolin 1981). Symbiosis in this case exists on a smaller and less effective scale, although fermentation in the human colon also produces the same major VFA group as the ruminant stomach. Dreher (1987:193) suggests that approximately 5% of the human caloric intake may be derived from the microbially-produced fatty acids. Wolin (1981:1467) also suggests that "VFA produced in the large intestine could provide appreciable metabolizable energy for humans whose diets regularly contain large amounts of plant fiber." In this sense, Eastwood and Brydon (1985:118) have suggested the humans are, in part "coprophagic albeit by absorption from the colon of short-chain fatty acids." The main point here is that humans are *capable* of metabolizing the products of microbial fermentation; they just are not as biologically effective at it when compared to the ruminants.

Microbial fermentation within a ruminant's forestomach produces a series of byproducts that serve the energy needs of the host. These by-

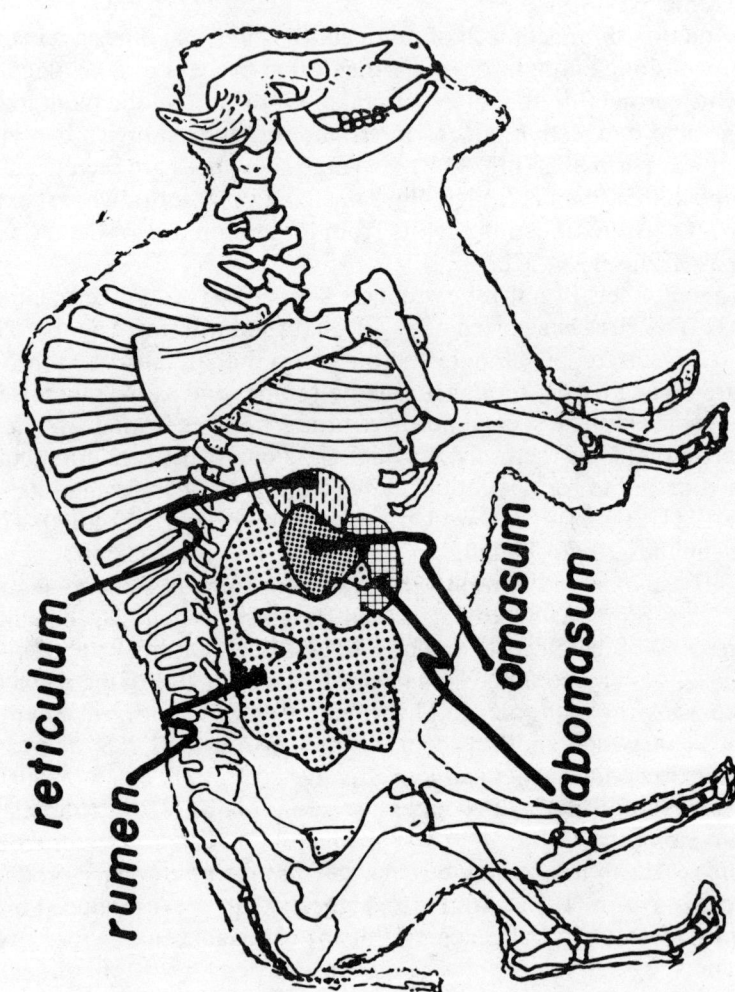

Figure 11.6. The four compartments of a ruminant stomach.

products could also be utilized by humans if they ate the rumen contents of prey along with the lean meat. It is significant that the stomach contents of ruminants comprise 15-20% of their total body weight at any given time. Ethnographic accounts of the use of "stomach contents" are fairly common, especially among arctic hunter-gatherers. It is interesting that the frequency of these occurrences appears to be higher among groups like the Caribou Eskimo, as well as among groups that do not fit the pattern of "effective temperature" (ET) to storage scale outlined by Binford (1980: 353). Accounts of these groups often highlight the use of the vegetable matter or plant foods in the stomachs of prey (e.g., Arima 1984:449; Murdock 1934:195; Savishinsky and Hara 1981:317). Stefansson (1926: 446) notes, for example, that the partly digested stomach contents of the Barren Ground caribou are frequently eaten frozen in the winter. Moreover, stomachs filled with reindeer moss were considered much better food than rumen contents from caribou that fed on the coarse, woody fibers of grassy plants. Similar accounts of human use of paunch contents exist for Plains groups (e.g., G. L. Wilson 1924:307), and Fry's identification of "large amounts of finely chopped shrub stems that are undoubtedly herbivore browse" (Fry 1970:249) in human coprolites from Hogup cave provides some archaeological evidence for consumption of herbivore stomach contents as well.

Common reference to the consumption of stomach contents has led to a misinterpretation of the role of "ruminophagism". The stomach contents of prey may have been more than a minimal source of secondary vegetable matter to the diet, because of their potential nutritive value is directly complementary to lean meat. Perhaps our models should also consider the byproducts of microbial fermentation in the form of VFA that "constitute the host's share in the carbohydrates of the food" (Smithers 1983:578), as well as other important dietary products such as the B vitamins (Dziuk 1984:333) produced in the ruminant stomach. Immediate consumption of ruminant stomach contents (while the products of fermentation were still nutritionally viable), and perhaps segments of the stomach wall into which nutrients were being absorbed, is not the same as eating plant fibers or grasses--the products of microbial fermentation produce a variety of additional energy sources in forms that are usable by humans.

CONCLUSION

The bison from the cold season Paleoindian bonebeds could have provided the full complement of proteins required by humans in the form of lean meat from the muscle tissues and carbohydrates from the fermentation products within the rumen. Since the stomach contents of ruminants comprise a substantial proportion of their total body weight, large quantities would have been available. If the length of seasonal availability of ruminant body fat were shorter, then there may have been no need for humans to store fats or carbohydrates in any form, since the period of physiological stress to human predators would also have been shorter and could have been alleviated with only minimal fat/carbohydrate inputs. Whether the by-products of microbial fermentation were indeed used by Paleoindians requires additional investigation. However, models of Pleistocene resource structure that do not account to the supplemental nutritive potential of prey stomach contents may produce a distorted view of Plains paleoecology in the meantime.

Under conditions of less accentuated seasonality coupled with a need for extensive residential mobility, use of a combination of protein from lean meat together with alternative food sources, such the products of microbial fermentation from the rumen, may have provided an important alternative to fall storage of fats. While only a speculative suggestion at this stage, reliable interpretations of hunter-gatherer foraging strategies on the basis of age profiles and seasons of death require their incorporation into more comprehensive models of regional and seasonal resource structure.

ACKNOWLEDGMENTS

Jack Fisher, Martha Graham, Jack Hofman, Rhoda Lewis, and Mark Sant all provided useful comments and criticisms of drafts of this paper and their efforts are greatly appreciated. Mary Stiner's editorial efforts have added considerable clarity to my presentation of the ideas in this paper.

References Cited

Agenbroad, L.
1978   *The Hudson-Meng Site: An Alberta Bison Kill in the Nebraska High Plains*. University Press of America, Washington, D. C.

Albanese, J. P.
1977   Paleotopography and Paleoindian Sites in Wyoming and Colorado. In Paleoindian Lifeways, edited by E. Johnson, pp. 28-34. *The Museum Journal* 17. West Texas Museum Association, Texas Tech University, Lubbock.

Altmann, J.
1974   Observational Study of Behavior: Sampling Methods. *Behavior* 48:1-41.

Altuna, J.
1986   The Mammalian Faunas From the Prehistoric Site of La Riera. In *La Riera Cave: Stone Age Hunter-gatherer Adaptations in Northern Spain*, edited by L. G. Straus and G. A. Clark, pp. 237-274 and 421-479 (Appendix B). Arizona State University Anthropological Research Paper, No. 36.

Alvard, M., and H. Kaplan
n.d.   Indigenous Human Predation on Neotropical Primates. Manuscript in preparation, in possession of authors, Department of Anthropology, University of New Mexico, Albuquerque.

Arima, E. Y.
1984   Caribou Eskimo. In *Handbook of North American Indians, Volume 5: Arctic*, edited by D. Damas, pp. 447-436. Smithsonian Institution Press, Washington, D. C.

Attwell, C. A. M.
1980   Age Determination of the Blue Wildebeest *Connochaetes taurinus* in Zululand. *South African Journal of Zoology* 15:121-130.

Bailey, G., and J. Parkington
1988   The Archaeology of Prehistoric Coastlines: An Introduction. In *The Archaeology of Prehistoric Coastlines*, edited by G. Bailey and J. Parkington, pp. 1-10. Cambridge University Press, Cambridge.

Ballard, W. B., and J. S. Whitman
1987   Marrow Fat Dynamics in Moose Calves. *Journal of Wildlife Management* 51:66-69.

Bamforth, D. B.
1988   *Ecology and Human Organization on the Great Plains.* Plenum Press, New York.

Banfield, A. W. F.
1974   *The Mammals of Canada.* University of Toronto Press, Toronto.

Barr, W.
1971   Postglacial Isostatic Movement in Northeastern Devon Island: A Reappraisal. *Arctic* 24:249-268.

Bearder, S. K.
1977   Feeding Habits of Spotted Hyaenas in a Woodland Habitat. *East African Wildlife Journal* 15:263-280.

Beckerman, S.
1980   Fishing and Hunting by the Bari of Colombia. In *Studies in Hunting and Fishing the Neotropics,* edited by R. Hames and K. Kensinger, pp. 67-111. Bennington College, Working Papers on South American Indians, No. 2, Bennington, Vermont.

Beckerman, S., and T. Sussenbach
1983   A Quantitative Assessment of the Dietary Contribution of Game Species to the Subsistence of South American Tropical Forest Tribal Peoples. In *Animals and Archaeology, I. Hunters and their Prey,* edited by J. Clutton-Brock and C. Grigson, pp. 337-350. BAR International Series 163.

Bedord, J. N.
1974   Morphological Variation in Bison Metacarpals and Metatarsals. In *The Casper Site: A Hell Gap Bison Kill on the High Plains,* edited by G. C. Frison, pp. 199-240. Academic Press, New York.

Behrensmeyer, A. K.
1975   The Taphonomy and Paleoecology of Plio-Pleistocene Vertebrate Assemblages East of Lake Rudolf, Kenya. *Bulletin of the Museum of Comparative Zoology, Harvard University,* No. 146:473-578.
1987   Taphonomy and Hunting. In *The Evolution of Human Hunting,* edited by M. H. Nitecki and D. V. Nitecki, pp. 423-450. Plenum Press, New York.

Behrensmeyer, A. K., and A. P. Hill (editors)
1980   *Fossils in the Making.* University of Chicago Press, Chicago.

Behrensmeyer, A. K., D. Western, and D. Dechant-Boaz
1979   New Perspectives in Vertebrate Paleoecology From a Recent Bone Assemblage. *Paleobiology* 5:12-21.

Benedict, J. B.
1975 The Murray Site: A Late Prehistoric Game Drive System in the Colorado Rocky Mountains. *Plains Anthropologist* 20(69):161-174.

Berger, J.
1986 *Wild Horses of the Great Basin, Social Competition and Population Size.* University of Chicago Press, Chicago.

Bergerud, A. T., and W. B. Ballard
1988 Wolf Predation on Caribou: The Nelchina Herd Case History, A Different Interpretation. *Journal of Wildlife Management* 52:344-357.

Bergerud, A. T., and J. B. Snider
1988 Predation in the Dynamics of Moose Populations: A Reply. *Journal of Wildlife Management* 52:559-564.

Bertram, B. C. R.
1979 Serengeti Predators and Their Social Systems. In, *Serengeti: Dynamics of an Ecosystem*, edited by A. R. E. Sinclair and M. Norton-Griffiths, pp. 221-248. University of Chicago Press, Chicago.

Bibikov, D. I.
1982 Wolf Ecology and Management in the USSR. In, *Wolves of the World: Perspectives of Behavior, Ecology, and Conservation*, edited by F. H. Harrington and P. C. Paquet, pp. 120-133. Noyes Publications, Park Ridge, N.J.

Bietti, A.
1976-77a The Excavation 1955-1959 in the Upper Palaeolithic Deposit of the Rockshelter at Palidoro (Rome, Italy). *Quaternaria* 19:149-155.
1976-77b Analysis and Illustration of the Epigravettian Industry Collected During the 1955 Excavations at Palidoro (Rome, Italy). *Quaternaria* 19:197-215.
1976-77c The Excavation 1955-1959 in the Upper Paleolithic Deposit of the Rock Shelter at Palidoro (Rome, Italy): A Brief General Introduction. *Quaternaria* 19:149-155.

Bietti, A., and M. C. Stiner
n.d. Subsistence and Settlement Patterns of the Italian Epigravettian: The Case of Riparo Salvini, Latium. In *Colioque International, Le Peuplement Magdalenien*, edited by H. Laville, J.-F. Rigaud, and B. Vandermeersch. Paris: Edition C.T.H.S., volume in preparation.

Bietti, A., G. Manzi, P. Passarello, A. G. Segre, and M. C. Stiner.
1988 The 1986 Excavation Campaign at Grotta Breuil (Monte Circeo LT). *Quaderni del Centro di Studio per l'Archeologia Etrusco-Italica, Archeologia Laziale* 9:372-388.

Bigg, M. A.
   1969   The Harbour Seal in British Columbia. *Bulletin of the Fisheries Research Board of Canada* 172.

Binford, L. R.
   1978   *Nunamiut Ethnoarchaeology.* Academic Press, New York.
   1980   Willow Smoke and Dogs' Tails: Hunter-gatherer Settlement Systems and Archaeological Site Formation. *American Antiquity* 45:4-20.
   1981   *Bones: Ancient Men and Modern Myths.* Academic Press, New York.
   1983   *In Pursuit of the Past.* Thames and Hudson, London.
   1984   *Faunal Remains from Klasies River Mouth.* Academic Press, New York.
   1985   Human Ancestors: Changing Views of Their Behavior. *Journal of Anthropological Archaeology* 4:292-327.

Binford, L. R., and J. B. Bertram
   1977   Bone Frequencies and Attritional Processes. In *For Theory Building in Archaeology,* edited by L. R. Binford, pp. 77-153. Academic Press, New York.

Blake, W., Jr.
   1975   Radiocarbon Age Determinations and Postglacial Emergence at Cape Storm, Southern Ellesmere Island, Arctic Canada. *Geografiska Annaler* 57(Series A):1-71.

Blalock, H. M., Jr.
   1979   *Social Statistics,* (Revised Second Edition). McGraw-Hill, New York.

Blanc, A. C., and A. G. Segre
   1953   Excursion au Mont Circé. *INQUA IVe Congres International* Roma, Pisa.

Blumenschine, R. J.
   1986a   Carcass Consumption Sequences and the Archaeological Distinction of Scavenging and Hunting. *Journal of Human Evolution* 15:639-659.
   1986b   *Early Hominid Scavenging Opportunities: Implications of Carcass Availability in the Serengeti and Ngorongoro Ecosystems.* BAR International Series 283.
   1987   Characteristics of an Early Hominid Scavenging Niche. *Current Anthropology* 28:383-407.
   1988a   Reinstating the Early Hominid Scavenging Niche: A Reply to Potts. *Current Anthropology* 29:483-486.
   1988b   An Experimental Model of the Timing of Hominid and Carnivore Influence on Archaeological Bone Assemblages. *Journal of Archaeological Science* 15:483-502.

References Cited 243

Blumenschine, R. J., and T. M. Caro
1986 Unit Flesh Weights of Some East African Bovids. *Journal of African Ecology* 24:273-286.

Bonner, W. N.
1979 Harbour (Common) Seal. In *Mammals in the Seas, Vol. II., Pinniped Summaries and Report on Sirenians*, pp. 58-62. Food and Agriculture Organization of the United Nations, Fisheries Series, No. 5.

Borden, C. E.
1975 Origins and Development of Early Northwest Coast Cultures to About 3000 B.C. *National Museum of Man Mercury Series, Archaeological Survey of Canada, Report*, No. 45.

Brain, C. K.
1981 *The Hunters or the Hunted?*. University of Chicago Press, Chicago.

Branan, W., and R. Marchinton
1987 Reproductive Ecology of White-tailed and Red Brocket Deer in Suriname. In *Ecology and Management of the Cervidae*, edited by C. Wemmer, pp. 344-351. Smithsonian Institution Press, Washington, D. C.

Breiwick, J. M., L. L. Eberhardt, and H. W. Braham
1984 Population Dynamics of Western Arctic Bowhead Whales (*Balaena mysticetus*). *Canadian Journal of Fisheries and Aquatic Sciences* 41:484-496.

Brink, J., and B. Dawe
1989 *Final Report of the 1985 and 1986 Field Season at Head-Smashed-In Buffalo Jump, Alberta*, No. 16. Archaeological Survey of Alberta, Edmonton.

Brink, J., M. Wright, B. Dawe, and D. Glaum
1986 Final Report on the 1984 Season at Head-Smashed-In Buffalo Jump Alberta. *Archaeological Survey of Alberta Manuscript Series*, No. 9. Archaeological Survey of Alberta, Edmonton.

Brooks, P. M.
1978 Relationship Between Body Condition and Age, Growth, Reproduction and Social Status in Impala, and Its Application to Management. *South African Journal of Wildlife Research* 8:151-157.

Brooks, P. M., J. Hanks, and J. V. Ludbrook
1977 Bone Marrow as an Index of Condition in African Ungulates. *South African Journal of Wildlife Research* 7:61-66.

Brueggeman, J. J.
 1982   Early Spring Distribution of Bowhead Whales in the Bering Sea. *Journal of Wildlife Management* 46:1036-1044.

Bunn, H. T.
 1981   Archaeological Evidence for Meat-eating by Plio-Pleistocene Hominids from Koobi Fora and Olduvai Gorge. *Nature* 291:574-577.
 1986   Patterns of Skeletal Representation and Hominid Subsistence Activities at Olduvai Gorge, Tanzania, and Koobi Fora, Kenya. *Journal of Human Evolution* 15:673-690.

Bunn, H. T., L. E. Bartram, and E. M. Kroll
 1988   Variability in Bone Assemblage Formation from Hadza Hunting, Scavenging, and Carcass Processing. *Journal of Anthropological Archaeology* 7:412-457.

Bunn, H. T., and R. J. Blumenschine
 1987   On "Theoretical Framework and Tests" of Early Hominid Meat and Marrow Acquisition: A Reply to Shipman. *American Anthropologist* 89:444-448.

Bunn, H. T., and E. M. Kroll
 1986   Systematic Butchery by Plio/Pleistocene Hominids at Olduvai Gorge, Tanzania. *Current Anthropology* 27:431-452.
 1988   Reply to Fact and Fiction About the Zinjanthropus Floor. *Current Anthropology* 29:135-149.

Butynski, T. M.
 1982   Vertebrate Predation by Primates: A Review of Hunting Patterns and Prey. *Journal of Human Evolution* 11:421-430.

Butzer, K. W.
 1978   Geoecological Perspectives on Early Hominid Evolution. In *Early Hominids of Africa*, edited by C. J. Jolly, pp. 191-217. St. Martin's Press, New York.

Carlson, C.
 1979   The Early Component at Bear Cove. *Canadian Journal of Archaeology* 3: 177-194.

Cassoli, P.
 1976-77   Upper Paleolithic Fauna at Palidoro (Rome): 1955 Excavations. *Quaternaria* 19:187-195.

Caughley, G.
 1966   Mortality Patterns in Mammals. *Ecology* 47:906-917.
 1977   *Analysis of Vertebrate Populations.* John Wiley and Sons, London.

Cavallo, J. A.
n.d.   Larders in the Limbs: Early Hominid Arboreal Scavenging Opportunities. Manuscript on file at the Department of Anthropology, Rutgers University.

Cavallo, J. A., and R. J. Blumenschine
1989   Tree-stored Leopard Kills: Expanding the Hominid Scavenging Niche. *Journal of Human Evolution* 18:393-399.

Chapman, D., and N. Chapman
1975   *Fallow Deer: Their History, Distribution and Biology*. Terence Dalton Ltd., Lavenham, Suffolk.

Chapman, C., L. Fedigan, and L. Chapman
1989   Post-weaning Resource Competition and Sex Ratios in Spider Monkeys. *Oikos* 54:315-319.

Charvon, E., and G. Orians
1973   Optimal Foraging: Some Theoretical Explorations. Unpublished manuscript, on file at the University of Utah, Salt Lake City, Utah.

Chomko, S. A., and B. M. Gilbert
1987   The Late Pleistocene/Holocene Faunal Record in the Northern Bighorn Mountains, Wyoming. In *Late Quaternary Mammalian Biogeography and Environments of the Great Plains and Prairies*, edited by R. W. Graham, H. A. Semken, Jr., and M. A. Graham, pp. 394-408. Illinois State Museum Scientific Papers, No. 22, Springfield.

Clark, J. G. D.
1946   Seal-hunting in the Stone Age of North-western Europe: A Study in Economic Prehistory. *Proceedings of the Prehistoric Society* 2:12-48.

Clutton-Brock, T. H., F. E. Guinness, and S. D. Albon
1982   *Red Deer: Behavior and Ecology of Two Sexes*. University of Chicago Press, Chicago.

Coe, M.
1980   The Role of Modern Ecological Studies in the Reconstruction of Paleoenvironments in Sub-Saharan Africa. In *Fossils in the Making*, edited by A. K. Behrensmeyer and A. P. Hill, pp. 55-67. University of Chicago Press, Chicago.

Coe, R. J., R. L. Downing, and B. S. McGinnes
1980   Sex and Age Bias in Hunter-killed White-tailed Deer. *Journal of Wildlife Management* 44:245-249.

Collett, S.
　1981　Population Characteristics of *Agouti paca* (Rodentia) in Colombia. *Publications of the Museum, Biological Series, Michigan State University* 5:489-601.

Conover, K.
　1978　Matrix Analysis. In *Studies in Bella Bella Prehistory*, edited by J. J. Hester and S. M. Nelson, pp. 67-99. Simon Fraser University, Department of Archaeology Publication No. 5.

Crandall, L. S.
　1965　Record of African Antelopes in the New York Zoological Park. *International Zoo Yearbook* 5:52-55.

Cribb, R. L. D.
　1987　The Logic of the Herd: A Computer Simulation of Archaeological Herd Structure. *Journal of Anthropological Archaeology* 6:376-415.

Crockett, C., and J. Eisenberg
　1987　Howlers: Variation in Group Size and Demography. In *Primate Societies*, edited by B. Smuts, D. Cheney, R. M. Seyfarth, R. Wrangham, and T. Strusaker, pp. 54-68. University of Chicago Press, Chicago.

Crouch, J. E.
　1972　*Functional Human Anatomy*. Lea and Febiger, Philadelphia.

Cubbage, J. C., and J. Calambokidis
　1987　Size-class Segregation of Bowhead Whales Discerned Through Aerial Stereophotogrammetry. *Marine Mammal Science* 3:179-185.

Dahl, G., and A. Hjort
　1976　*Having Herds. Pastoral Herd Growth and Household Economy*. Department of Anthropology, University of Stockholm, Stockholm.

Davis, J. L., P. Valkenburg, and D. J. Reed
　1987　Correlations and Depletion Patterns of Marrow Fat in Caribou Bones. *Journal of Wildlife Management* 51:365-371.

Davis, R. A., and W. R. Koski
　1980　Recent Observations of Bowhead Whales in the Eastern Canadian High Arctic. *Report of the International Whaling Commission* 30:439-444.

Deniz, E., and S. Payne
　1982　Eruption and Wear in the Mandibular Dentition as a Guide to Aging Turkish Angora Goats. In *Aging and Sexing Animal Bones From Archaeological Sites*, edited by B. Wilson, C. Grigson, and S. Payne, pp. 155-205. BAR British Series 109, Oxford.

References Cited

Diamond, J. M.
1975 The Assembly of Species Communities. In *Ecology and Evolution of Communities*, edited by M. L. Cody and J. M. Diamond, pp. 342-444. Belknap Press.
1989 Quaternary Megafaunal Extinctions: Variations on a Theme by Paganini. *Journal of Archaeological Science* 16:167-175.

Dreher, M. L.
1987 *The Handbook of Dietary Fiber: An Applied Approach*. M. Dekker, New York.

Dubost, G.
1980 L'Ecologie et la Vie Sociale du Cephalophe Bleu (*Cephalophus monticola*, Thurnberg), Petit Ruminant Forestier Africain. *Zeitschrift fur Tierpsychologie/Journal of Comparative Ethology* 54:205-266.

Ducos, P.
1968 L'Origine des Animaux Domestiques en Palestine. *L'Institut de Préhistoire de l'Université de Bordeaux Mémoire* 6.

Dwyer, P. D.
1974 The Price of Protein: Five Hundred Hours of Hunting in the New Guinea Highlands. *Oceania* 44:278-293.

Dyke, A. S.
1979 Radiocarbon-dated Holocene Emergence of Somerset Island, Central Canadian Arctic. In *Current Research, Part B. Geological Survey of Canada, Paper 79-1B*:307-318.
1980 Redated Holocene Whale Bones From Somerset Island, District of Franklin. In *Current Research, Part B, Geological Survey of Canada*, Paper 80-1B: 169-270.

Dyke, A. S., and T. F. Morris
1990 Postglacial History of the Bowhead Whale and of Driftwood Penetration: Implications for Paleoclimate, Central Canadian Arctic. *Geological Survey of Canada, Paper*:89-124.

Dziuk, H. E.
1984 Digestion in the Ruminant Stomach. In *Dukes' Physiology of Domestic Animals* (Tenth Edition), edited by M. J. Swenson, pp. 320-339. Cornell University Press, Ithaca.

Eastwood, M., and W. G. Brydon
1985 Physiological Effects of Dietary Fibre on the Alimentary Tract. In *Dietary Fibre, Fibre Depleted Foods and Disease*, edited by H. Trowell, D. Burkitt, and K. Heaton, pp. 105-131. Academic Press, London.

Eisenberg, J. F.
    1981   *The Mammalian Radiations: An Analysis of Trends in Evolution, Adaptation, and Behavior.* University of Chicago Press, Chicago.
    1989   *Mammals of the Neotropics: The Northern Neotropics.* University of Chicago Press, Chicago.

Egan, Maj. H.
    1917   *Pioneering the West,* edited by W. M. Egan. Salt Lake City.

Elder, W. H.
    1965   Primeval Deer Hunting Pressures Revealed by Remains From Indian Middens. *Journal of Wildlife Management* 29:366-370.

Emerson, T. E.
    1980   A Stable White-tailed Deer Population Model and Its Implications for Interpreting Prehistoric Hunting Patterns. *Mid-Continental Journal of Archaeology* 5:117-132.

Emmons, L.
    1990   *Neotropical Rainforest Mammals.* University of Chicago Press, Chicago.

Erickson, J. A., and W. G. Seliger
    1969   Efficient Sectioning of Incisors for Estimating Ages of Mule Deer. *Journal of Wildlife Management* 33:384-388.

Ewbank, J. M., D. W. Phillipson, and R. D. Whitehouse (with E. S. Higgs)
    1964   Sheep in the Iron Age: A Method of Study. *Proceedings of the Prehistoric Society* 30:423-426.

Ewer, R. F.
    1973   *The Carnivores.* Weidenfeld and Nicholson, London.

Ewers, J. C.
    1955   The Horse in Blackfoot Indian Culture. *Bureau of American Ethnology Bulletin* 159.

Fawcett, W. B., Jr.
    1987   *Communal Hunts, Human Aggregations, Social Variation, and Climatic Change: Bison Utilization by Prehistoric Inhabitants of the Great Plains.* Unpublished Ph.D. dissertation, Department of Anthropology, University of Massachusetts, Amherst.

Fisher, D. C.
    1987   Mastodont Procurement by Paleoindians of the Great Lakes Region: Hunting or Scavenging? In *The Evolution of Human Hunting,* edited by M. H. Nitecki and D. V. Nitecki, pp. 309-421. Plenum Press, New York.

References Cited

Foley, R.
1982 Reconsideration of the Role of Predation on Large Mammals in Tropical Hunter-gatherer Adaptation. *Man* 17:393-402.

Fox, M. W.
1984 *The Whistling Hunters: Field Studies of the Asiatic Wild Dog (Cuon alpinus).* SUNY Press, Albany.

Frame, G. W.
1986 Carnivore Competition and Resource Use in the Serengeti Ecosystem of Tanzania. Unpublished Ph.D. dissertation, Utah State University, Logan.

Freeman, M. M. R.
1979 A Critical View of Thule Culture and Ecological Adaptation. *In* Thule Eskimo Culture: An Anthropological Retrospective, edited by A. P. McCartney, pp. 278-285. *National Museum of Man Mercury Series, Archaeological Survey of Canada Paper,* No. 88.

Frison, G. C.
1967 The Piney Creek Sites, Wyoming (48JO311 and 312). *University of Wyoming Publications* 33:1-92. University of Wyoming, Laramie.
1970 The Glenrock Buffalo Jump, 48CO304: Late Prehistoric Period Buffalo Procurement and Butchering. *Plains Anthropologist Memoir* No. 7.
1971 The Buffalo Pound in Northwestern Plains Prehistory: Site 48CA302, Wyoming. *American Antiquity* 36:77-91.
1978a Animal Population Studies and Cultural Inference. *In* Bison Procurement and Utilization: A Symposium, edited by L. B. Davis, and M. Wilson, pp. 44-52. *Plains Anthropologist Memoir* 14.
1978b *Prehistoric Hunters of the High Plains.* Academic Press, New York.
1982a Bison Dentition Studies. In *The Agate Basin Site: A Record of the Paleoindian Occupation of the Northwestern High Plains,* edited by G. C. Frison and D. Stanford, pp. 240-260. Academic Press, New York.
1982b Folsom Components. In *The Agate Basin Site: A Record of the Paleoindian Occupation of the Northwestern High Plains,* edited by G. C. Frison and D. Stanford, pp. 37-76. Academic Press, New York.
1982c Hell Gap Components. In *The Agate Basin Site: A Record of the Paleoindian Occupation of the Northwestern High Plains,* edited by G. C. Frison and D. Stanford, pp. 135-142. Academic Press, New York.
1982d Paleo-Indian Winter Subsistence Strategies on the High Plains. *In* Plains Indian studies: A Collection of Essays in Honor of J. C. Ewers and Waldo Wedel, edited by D. H. Ubelaker and H. J. Viola, pp. 193-201. *Smithsonian Contributions to Anthropology* No. 30.
1984 The Carter/Kerr-McGee Paleoindian Site: Cultural Resource Management and Archaeological Research. *American Antiquity* 49:288-314.
1986 *The Colby Mammoth Site.* University of New Mexico Press, Albuquerque.

1987   Prehistoric, Plains-mountain, Large Mammal, Communal Hunting Strategies. In *The Evolution of Human Hunting*, edited by M. H. Nitecki and D. V. Nitecki, pp. 177-223. Plenum, New York.
1989   Experimental Use of Clovis Weaponry and Tools on African Elephants. *American Antiquity* 54:766-784.

Frison, G. C., R. L. Andrews, J. M. Adovasio, R. C. Carlisle, and R. Edgar
1986   A Late Paleoindian Animal Trapping Net From Northern Wyoming. *American Antiquity* 51:352-361.

Frison, G. C., and C. A. Reher
1970   Age Determination of Buffalo by Teeth Eruption and Wear. *In* The Glenrock Buffalo Jump 48CO304, edited by G. C. Frison, Appendix I. *Plains Anthropologist Memoir* No. 7.

Frison, G. C., and L. C. Todd
1986   *The Horner Site: The Type Site of the Cody Cultural Complex*. Academic Press, New York.

Frison, G. C., M. Wilson, and D. N. Walker
1978   The Big Goose Creek Site: Bison Procurement and Faunal Analysis. *Occasional Papers on Wyoming Archaeology* No. 1. Department of Anthropology, Laramie.

Frison, G. C., M. Wilson, and D. J. Wilson
1976   Fossil Bison and Artifacts from a Early Altithermal Period Arroyo Trap in Wyoming. *American Antiquity* 41:28-57.

Fry, G. F.
1970   Preliminary Analysis of the Hogup Cave Coprolites. *In* Hogup Cave, by C. M. Aikens, pp. 247-250. *University of Utah Anthropological Papers* No 93. University of Utah Press, Salt Lake.

Fulgham, T., and D. Stanford
1982   The Frasca Site: A Preliminary Report. *Southwestern Lore* 48:1-9.

Fuller, T. K., and Ll. B. Keith
1980   Wolf Predation Dynamics and Prey Relationships in Northeastern Alaska. *Journal of Wildlife Management* 44:583-602.

Fuller, T. K., P. L. Coy, and W. J. Peterson
1986   Marrow Fat Relationships Among Leg Bones of White-Tailed Deer. *Wildlife Society Bulletin* 14:73-75.

Gamble, C.
1986 *The Palaeolithic Settlement of Europe*. Cambridge University Press, Cambridge.
1987 Man the Shoveler: Alternative Models for Middle Pleistocene Colonization and Occupation in Northern Latitudes. In *The Pleistocene Old World: Regional Perspectives*, edited by O. Soffer, pp. 81-98. Plenum Press, New York.

Gavin, T. A., L. H. Suring, P. A. Vohs, Jr., and E. C. Meslow
1984 Population Characteristics, Spatial Organization, and Natural Mortality in the Columbian White-tailed Deer. *Wildlife Monographs* 91.

Geist, V.
1971 *Mountain Sheep: A Study of Behavior and Evolution*. University of Chicago Press, Chicago.

Gentry, R. L., and D. E. Withrow
1978 Steller Sea Lion. In *Marine Mammals*, edited by D. Haley, pp. 167-171. Pacific Search Press, Seattle.

Gifford, D. P.
1981 Taphonomy and Paleoecology: A Critical Review of Archaeology's Sister Disciplines. *Advances in Archaeological Method and Theory* 4:365-438.

Gifford-Gonzalez, D.
1989 Shipman's Shaky Foundations. *American Anthropologist* 91:180-186.
n.d.a Evaluating the Quadratic Crown Height Aging Method With Known-age Bison Samples. Manuscript submitted to Journal of Archaeological Science.
n.d.b Regressions for Aging Bovine Molars Using Remnant Crown Height. Manuscript in preparation, on file at Board of Studies in Anthropology, Univesity of California, Santa Cruz.

Gladwin, H. S.
1947 *Men Out of Asia*. McGraw-Hill, New York.

Gordon, E.
n.d. Evaluating Dental Age Profiles in Blue Duikers. Manuscript on file at the Department of Anthropology, University of California, Santa Barbara.

Gould, R. A., and P. J. Watson
1982 A Dialogue on the Meaning and Use of Analogy in Ethnoarchaeological Reasoning. *Journal of Anthropological Archaeology* 1:355-381.

Graham, M. A., M. C. Wilson, and R. W. Graham
1987 Paleoenvironments and Mammalian Faunas of Montana, Southern Alberta, and Southern Saskatchewan. In *Late Quaternary Mammalian Biogeography and Environments of the Great Plains and Prairies*, edited by R. W. Graham,

H. A. Semken, Jr., and M. A. Graham, pp. 410-459. Illinois State Museum Scientific Papers, No. 22, Springfield.

Graham, R. W.
1986 Plant-animal Interactions and Pleistocene Extinctions. In *Dynamics of Extinctions*, edited D. K. Elliot, pp. 131-154. J. Wiley, New York.
1987 Late Quaternary Mammalian Faunas and Paleoenvironments of the Southwestern Plains of the United States. In *Late Quaternary Mammalian Biogeography and Environments of the Great Plains and Prairies*, edited by R. W. Graham, H. A. Semken, Jr., and M. A. Graham, pp. 24-86. Illinois State Museum Scientific Papers, No. 22, Springfield.

Graham, R. W., and E. L. Lundelius, Jr.
1984 Coevolutionary Disequilibrium and Pleistocene Extinctions. In *Quaternary Extinctions: A Prehistoric Revolution*, edited by P. S. Martin and R. G. Klein, pp. 223-249. University of Arizona Press, Tucson.

Grant, A.
1982 The Use of Tooth Wear as a Guide to the Age of Domestic Animals. In *Aging and Sexing Animal Bones from Archaeological Sites*, edited by B. Wilson, C. Grigson and S. Payne, pp. 91-108. BAR British Series 109, Oxford.

Grayson, D. K.
1973 On the Methodology of Faunal Analysis. *American Antiquity* 38:432-439.
1989 Bone Transport, Bone Destruction, and Reverse Utility Curves. *Journal of Archaeological Science* 16:643-652.

Grigson, C.
1982 Sex and Age Determination of Some Bones and Teeth of Domestic Cattle: A Review of the Literature. In *Aging and Sexing Animal Bones From Archaeological Sites*, edited by B. Wilson, C. Grigson, and S. Payne, pp. 7-23. BAR British Series 109, Oxford.

Grønnow, B.
1987 Meiendorf and Stellmoor Revisited: An Analysis of Late Palaeolithic Reindeer Exploitation. *Acta Archaeologica* 56(1985):131-166. Copenhagen, Denmark.

Gusinde, M.
1961 *The Yamana* (translated from the German by F. Schutze), Volumes 1-5 (1918-24). HRAF, New Haven.

Guthrie, R. D.
1980 Bison and Man in North America. *Canadian Journal of Anthropology* 1:55-73.

1984   Mosaics, Allelochemics and Nutrients: An Ecological Theory of Late Pleistocene Megafaunal Extinctions. In *Quaternary Extinctions: A Prehistoric Revolution*, edited by P. S. Martin and R. G. Klein, pp. 259-298. University of Arizona Press, Tucson.

Hames, R.
1979   A Comparison of the Efficiencies of the Shotgun and the Bow in Neotropical Forest Hunting. *Human Ecology* 7:219-252.

Hames, R. B., and W. T. Vickers
1982   Optimal Diet Breadth Theory as a Model to Explain Variability in Amazonian Hunting. *American Ethnologist* 9:358-378.

Hasegawa, T., M. Hiraiwa, T. Nishida, and H. Takasaki
1983   New Evidence on Scavenging Behavior in Wild Chimpanzees. *Current Anthropology* 24:231-232.

Hauge, T. M., and Ll. B. Keith
1981   Dynamics of Moose Populations in Northeastern Alaska. *Journal of Wildlife Management* 45:573-597.

Haury, E. W., E. B. Sayles, and W. W. Wasley
1959   The Lehner Mammoth Site, Southeastern Arizona. *American Antiquity* 25:2-30.

Haynes, G.
1980   Prey Bones and Predators: Potential Ecologic Information from Analysis of Bone Sites. *Ossa* 7:75-97.
1982   Utilization and Skeletal Disturbances of North American Prey Carcasses. *Arctic* 35:266-281.
1985   Age Profiles in Elephant and Mammoth Bone Assemblages. *Quaternary Research* 24:333-345.
1987   Proboscidean Die-offs and Die-outs: Age Profiles in Fossil Collections. *Journal of Archaeological Science* 14:659-668.
1988a   Longitudinal Studies of African Elephant Death and Bone Deposits. *Journal of Archaeological Science* 15:131-57.
1988b   Mass Death and Serial Predation: Comparative Taphonomic Studies of Modern Large Mammal Death Sites. *Journal of Archaeological Science* 15: 219-235.

Hecker, H. M.
1982   Domestication Revisited: Its Implications for Faunal Analysis. *Journal of Field Archaeology* 9:217-236.

Henschel, J. R., R. Tilson, and F. von Blottnitz
   1979   Implications of a Spotted Hyaena Bone Assemblage in the Namib Desert. *South African Archaeological Bulletin* 34:127-131.

Higgs, E. S., and M. R. Jarman
   1972   The Origins of Animal and Plant Husbandry. In *Papers in Economic Prehistory*, edited by E. S. Higgs, pp. 3-13. Cambridge University Press, Cambridge.

Hildebrandt, W. R.
   1984   Late Period Hunting Adaptations on the North Coast of California. *Journal of California and Great Basin Anthropology* 6:189-206.

Hill, K., and K. Hawkes
   1983   Neotropical Hunting Among the Ache of Eastern Paraguay. In *Adaptive Responses of Native Amazonians*, edited by R. Hames and W. Vickers, pp. 139-188. Academic Press, New York.

Hill, K., H. Kaplan, K. Hawkes, and A. M. Hurtado
   1987   Foraging Decisions Among Ache Hunter-gatherers: New Data and Implications for Optimal Foraging Models. *Ethology and Sociobiology* 8:1-36.

Hillerud, J. M.
   1970   *Subfossil High Plains Bison*. Unpublished Ph.D. dissertation, Geology Department, University of Nebraska, Lincoln.

Hofman, J. L., L. C. Todd, and C. B. Schultz
   1990   The Lipscomb Bison Quarry: Continuing Investigations at a Folsom Kill-Butchery Site on the Southern Plains. *Bulletin of the Texas Archaeological Society* Volume 60. In press.

Holdridge, L. R.
   1967   *Life Zone Ecology: San José*. Tropical Research Center, Costa Rica.

Houston, D. C.
   1979   The Adaptations of Scavengers. In *Serengeti: Dynamics of an Ecosystem*, edited by A. R. E. Sinclair and M. Norton-Griffiths, pp. 263-286. University of Chicago Press, Chicago.

Hudson, J.
   1990a   Advancing Methods in Zooarchaeology, An Ethnoarchaeological Study Among the Aka Pygmies. Unpublished Ph.D. dissertation, Department of Anthropology, University of California, Santa Barbara.
   1990b   Identifying Food Sharing Archaeologically, An Ethnoarchaeological Approach. Paper presented at the Sixth International Conference on Hunting and Gathering Societies, Fairbanks.

Hungate, R. E.
   1966   *The Rumen and Its Microbes.* Academic Press, London.
   1984   Microbes of Nutritional Importance in the Alimentary Tract. *Proceedings of the Nutrition Society* 43:1-11.

Isaac, G. Ll.
   1983   Bones in Contention: Competing Explanations for the Juxtaposition of Early Pleistocene Artifacts and Faunal Remains. In *Animals and Archaeology, I. Hunters and their Prey,* edited by J. Clutton-Brock and C. Grigson, pp. 3-19. BAR International Series 163.

Isaac, G. Ll., and D. C. Crader
   1981   To What Extent Were Early Hominids Carnivorous? An Archaeological Perspective. In *Omnivorous Primates: Gathering and Hunting in Human Evolution,* edited by R. S. O. Harding and G. Teleki, pp. 37-103. Columbia University Press, New York.

Izawa, K.
   1976   Group Sizes and Compositions of Monkeys in the Upper Amazon Basin. *Primates* 17:367-399.

Jarman, M. R., and P. F. Wilkinson
   1972   Criteria of Animal Domestication. In *Papers in Economic Prehistory,* edited by E. S. Higgs, pp. 83-96. Cambridge University Press, Cambridge.

Jennings, D. C., and J. Hebbring
   1983   *Buffalo Management and Marketing.* National Buffalo Association, Custer, South Dakota.

Jobson, R. W., and W. R. Hildebrandt
   1980   The Distribution of Oceangoing Canoes on the North Coast of California. *Journal of California and Great Basin Anthropology* 2:165-174.

Jodry, M. A.
   1987   *Stewart's Cattle Guard Site: A Folsom Site in Southern Colorado: A Report of the 1981 and 1983 Field Season.* Unpublished Masters thesis, Department of Anthropology, University of Texas at Austin.

Johnson, E.
   1986   Late Pleistocene and Early Holocene Vertebrates and Paleoenvironments on the Southern High Plains, U.S.A. *Geographie Physique et Quaternaire* 40: 249-261.
   1987   (editor) *Lubbock Lake: Late Quaternary Studies on the Southern High Plains.* Texas A & M Press, College Station.

Johnson, E., and V. T. Holliday
  1980 Plainview Kill/Butchering Locale on the Llano Estacado — The Lubbock Lake Site. *Plains Anthropologist* 88:89-111.
  1981 Late Paleo-Indian Activity at the Lubbock Lake site. *Plains Anthropologist* 26(93):173-193.

Johnson, R. A., S. W. Carothers, and T. J. McGill
  1987 Demography of Feral Burros in the Mohave Desert. *Journal of Wildlife Management* 51:916-920.

Kaplan, H., and K. Kopischke
  1991 Resource Use, Traditional Technology and Change Among Native Peoples of Lowland South America. In *Conservation of Neotropical Forests: Building on Traditional Resource Use*, edited by K. Redford and C. Padoch (in press). Columbia University Press, New York.

Kehoe, T. F.
  1970 The Boarding School Bison Drive Site. *Plains Anthropologist, Memoir* No. 4.

Kelly, R. L., and L. C. Todd
  1988 Coming Into the Country: Early Paleoindian Hunting and Mobility. *American Antiquity* 53:231-244.

Kiltie, R. A., and J. Terborgh
  1983 Observations of the Behavior of Rain Forest Peccaries in Peru: Why Do White-lipped Peccaries Form Herds? *Z. Tierpsychol.* 62:241-255.

King, G. E.
  1975 Socioterritorial Units Among Carnivores and Early Hominids. *Journal of Anthropological Research* 31:69-87.

Klein, R. G.
  1973 *Ice-Age Hunters of the Ukraine*. University of Chicago Press, Chicago.
  1978 Stone Age Predation of Large African Bovids. *Journal of Archaeological Science* 5:195-217.
  1979 Stone Age Exploitation of Animals in Southern Africa. *American Scientist* 67:151-160.
  1981a Ungulate Mortality and Sedimentary Facies in the Late Tertiary Varswater Formation, Langebaanweg, South Africa. *Annals of the South African Museum* 84:233-254.
  1981b Stone Age Predation on Small African Bovids. *South African Archaeological Bulletin* 36:55-65.
  1982a Age (Mortality) Profiles as a Means of Distinguishing Hunted Species from Scavenged Ones in Stone Age Archaeological Sites. *Paleobiology* 8:151-158.
  1982b Patterns of Ungulate Mortality and Ungulate Mortality Profiles from Langebaanweg (Early Pliocene) and Elandsfontein (Middle Pleistocene), South-

western Cape Province, South Africa. *Annals of the South African Museum* 90:49-94.

1986 Review of L. R. Binford's "Faunal Remains From Klasies River Mouth". *American Anthropologist* 88:494-495.

1987 Reconstructing How Early People Exploited Animals: Problems and Prospects. In *The Evolution of Human Hunting*, edited by M. H. Nitecki and D. V. Nitecki, pp. 11-45. Plenum Press, New York.

1989 Why Does Skeletal Part Representation Differ Between Smaller and Larger Bovids at Klasies River Mouth and Other Archaeological Sites? *Journal of Archaeological Science* 16:363-381.

Klein, R. G., and K. Cruz-Uribe
1983a Age (Mortality) Profiles as a Means of Distinguishing Hunted Species From Scavenged Ones in Stone Age Archaeological Sites. *Paleobiology* 8:151-158.
1983b The Computation of Ungulate Age (Mortality) Profiles from Dental Crown Heights. *Paleobiology* 9:70-78.
1984 *The Analysis of Animal Bones from Archeological Sites*. University of Chicago Press, Chicago.

Klein, R. G., C. Wolf, L. G. Freeman, and K. Allwarden
1981 The Use of Dental Crown Heights for Constructing Age Profiles of Red Deer and Similar Species in Archaeological Samples. *Journal of Archaeological Science* 8:1-31.

Klein, R. G., K. Alwarden, and C. Wolf
1983 The Calculation and Interpretation of Ungulate Age Profiles from Dental Crown Heights. In *Hunter-gatherer Economy in Prehistory: A European Perspective*, edited by G. Bailey, pp. 47-57. Cambridge University Press, Cambridge.

Koike, H., and N. Ohtaishi
1985 Prehistoric Hunting Pressure Estimated by the Age Composition of Excavated Sika Deer (*Cervus nippon*) Using the Annual Layer of Tooth Cement. *Journal of Archaeological Science* 12:443-456.
1987 Estimation of Prehistoric Hunting Rates Based on the Age Composition of Sika Deer (*Cervus nippon*). *Journal of Archaeological Science* 14:251-269.

Kornfeld, M.
1988 The Rocky Folsom Site: A Small Folsom Assemblage From the Northwestern Plains. *North American Archaeologist* 9:197-222.

Kroeber, A. L., and S. A. Barrett
1960 Fishing Among the Indians of Northwestern California. *University of California Anthropological Records* 21:1-210.

Kruuk, H.
    1972   *The Spotted Hyaena.* University of Chicago Press, Chicago.

Kuchikuru, Y.
    1988   Efficiency and Focus of Blowpipe Hunting Among Semaq Beri Hunter-gatherers of Peninsular Malaysia. *Human Ecology* 16:271-305.

Kuhn, S. L.
    1989   Projectile Weapons and Investment in Food Procurement Technology in the Eurasian Middle Paleolithic (Abstract). *American Journal of Physical Anthropology* 78:257.
    1990a   Diversity Within Uniformity: Tool Manufacture and Use in the Pontinian Mousterian of Latium (Italy). Unpublished Ph.D. dissertation, University of New Mexico, Albuquerque.
    1990b   Late Mousterian Technology and Foraging Patterns at Grotta Breuil, Italy. *AnthroQuest* 42:12-15.

Kuhn, S. L., and M. C. Stiner
    n.d.   Bones and Stones: Foraging Practices, Land Use and Technology in the Italian Mousterian. In *The Organization of Land and Space Use and Technology,* edited by L. R. Binford, J. Piper and R. Kneebone, University of New Mexico, Albuquerque. Volume in press.

Kurtén, B.
    1953   On the Variation and Population Dynamics of Fossil and Recent Mammal Populations. *Acta Zoologica Fennica* 76:1-222.
    1971   *The Age of Mammals.* Weidenfeld and Nicolson, London.

Laj Pannocchia, F.
    1950   L'Industria Pontiniana della Grotta di S. Agostino (Gaeta). *Rivista di Scienze Preistoriche* 5:67-86

Lawson, J. W., and D. Renouf
    1987   Bonding and Weaning in Harbor Seals, *Phoca vitulina. Journal of Mammalogy* 68:445-449.

LeBeouf, F. J., and K. T. Briggs
    1977   The Cost of Living in a Seal Harem. *Mammalia* 41:167-195.

Ledger, H. P.
    1968   Body Composition as a Basis for a Comparative Study of Some East African Mammals. *Symposia of the Zoological Society of London* 21:289-310.

Lee, R. B., and I. DeVore
    1968   *Man the Hunter.* Aldine, Chicago.

Leslie, D. M., Jr., and C. L. Douglas
   1979   Desert Bighorn Sheep of the River Mountains, Nevada. *Wildlife Monographs,* No. 66.

Levine, M. A.
   1983   Mortality Models and the Interpretation of Horse Population Structure. In *Hunter-Gatherer Economy in Prehistory: A European Perspective,* edited by G. Bailey, pp. 23-46. Cambridge University Press, London.

Lindsay, D. G.
   1975   *Sea Ice Atlas of Arctic Canada 1961-1968.* Department of Energy, Mines and Resources, Canada. Ottawa.
   1977   *Sea Ice Atlas of Arctic Canada 1969-1973.* Department of Energy, Mines and Resources, Canada. Ottawa.
   1981   *Sea Ice Atlas of Arctic Canada 1975-1978.* Department of Energy, Mines and Resources, Canada. Ottawa.

Lorrain, D.
   1968   Analysis of the Bison Bone. *In* Bonfire Shelter: A Stratified Bison Kill Site, Val Verde County, Texas, edited by D. S. Dibble and D. Lorrain, pp. 77-132. *Texas Memorial Museum, Miscellaneous Paper* No. 1.

Lowe, V. P. W.
   1967   Teeth as Indicators of Age With Special Reference to Red Deer (*Cervus elaphus*) of Known Age from Rhum. *Journal of Zoology, London* 152:137-153.

Lyman, R. L.
   1982   Archaeofaunas and Subsistence Studies. In *Advances in Archaeological Method and Theory, Vol. 5,* edited by M. B. Schiffer, pp. 331-393. Academic Press, New York.
   1984   Bone Density and Differential Survivorship of Fossil Classes. *Journal of Anthropological Archaeology* 3:259-299.
   1985   Bone Frequencies: Differential Transport, In Situ Destruction, and the MGUI. *Journal of Archaeological Science* 12:221-236.
   1987a Hunting for Evidence of Plio-Pleistocene Hominid Scavengers. *American Anthropologist* 89:710-715.
   1987b On the Analysis of Vertebrate Mortality Profiles: Sample Size, Mortality Type, and Hunting Pressure. *American Antiquity* 52:125-142.
   1988   Zoogeography of Oregon Coast Marine Mammals: The Last 3000 Years. *Marine Mammal Science* 4:247-264.
   1989   Seal and Sea Lion Hunting: A Zooarchaeological Study from the Southern Northwest Coast of North America. *Journal of Anthropological Archaeology* 8:68-99.

Lyman, R. L., L. A. Clark, and R. E. Ross
    1988   Harpoon Stone Tips and Sea Mammal Hunting on the Oregon and Northern California Coast. *Journal of California and Great Basin Anthropology* 10: 73-87.

MacArthur, R. H.
    1968   The Theory of Niche. *Population Biology and Evolution,* edited by R. C. Lewontin, pp. 159-176. Syracuse University Press, Syracuse.
    1970   Species Packing and Competitive Equilibrium for Many Species. *Theoretical Population Biology* 1:1-11.

MacArthur, R. H., and R. Levins
    1967   The Limiting Similarity, Convergence, and Divergence of Coexisting Species. *The American Naturalist* 101:377-385.

Maher, W. J., and N. J. Wilimovsky
    1963   Annual Catch of Bowhead Whales by Eskimos at Point Barrow, Alaska, 1928-1960. *Journal of Mammalogy* 44:16-20.

Mansfield, A. W.
    1971   Occurrences of the Bowhead or Greenland Right Whale (*Balaena mysticetus*) in Canadian Arctic Waters. *Journal of the Fisheries Research Board of Canada* 28:1873-1875.

Marks, S. A.
    1976   *Large Mammals and a Brave People*. University of Washington Press, Seattle.

Marquette, W. M.
    1976   Bowhead Whale Field Studies in Alaska 1975. *Marine Fisheries Review* 38:9-17.
    1978   Bowhead Whale. In *Marine Mammals of the Eastern North Pacific and Arctic Waters,* edited by D. Haley, pp. 71-81. Pacific Search Press.

Marshall, F.
    1986   Implications of Bone Modification in a Neolithic Faunal Assemblage for the Study of Early Hominid Butchery and Subsistence Practices. *Journal of Human Evolution* 15:661-672.

Martin, L. D., and J. B. Martin
    1984   The Effect of Pleistocene and Recent Environments on Man in North America. *Current Research in the Pleistocene* 1:73-75.
    1986   Equability in the Late Pleistocene. In *Guidebook to the Friends of the Pleistocene,* edited by W. Johnson, pp. 123-127. State Geological Survey of Kansas, Lawrence.

Martin, P.
    1967   Pleistocene Overkill. *Natural History* 76:32-38.

Mate, B. R.
    1981   Marine Mammals. In *Natural History of Oregon Coast Mammals*, by C. Maser, B. R. Mate, J. F. Franklin and C. T. Dyrness, pp. 372-458. U.S.D.A. Forest Service General Technical Report PNW-133.

Mate, B. R., and R. L. Gentry
    1979   Northern (Steller) Sea Lion. In *Mammals in the Seas, Vol. II, Pinniped Species Summaries and Report on Sirenians*, pp. 1-4. Food and Agriculture Organization of the United Nations, Fisheries, No. 5.

Mathiassen, T.
    1927   Archaeology of the Central Eskimos. *Report of the Fifth Thule Expedition 1921-24*, Vol. 4.

Matson, R. G.
    1983   Intensification and the Development of Cultural Complexity: The Northwest Versus the Northeast Coast. In *The Evolution of Maritime Cultures on the Northeast and the Northwest Coasts of America*, edited by R. J. Nash, pp. 125-148. Simon Fraser University, Department of Archaeology Publication No. 11.

Maxwell, M. S.
    1985   *Prehistory of the Eastern Arctic*. Academic Press, Orlando.

McCartney, A. P.
    1977   Thule Eskimo Prehistory Along Northwestern Hudson Bay. *National Museum of Man Mercury Series*, No. 70.
    1978   *Study of Whale Bones for the Reconstruction of Canadian Arctic Bowhead Whale Stocks and Whale Use by Prehistoric Inuit*. Report Submitted to Northern Environmental Protection Branch, IANA, Ottawa.
    1979   Archaeological Whale Bone: A Northern Resource. *University of Arkansas Anthropological Papers*, No. 1.
    1980a   *Study of Archaeological Whale Bones for the Reconstruction of Canadian Arctic Bowhead Whale Stocks and Whale Use by Prehistoric Inuit*. Final Report Submitted to Northern Environmental Protection Branch, IANA, Ottawa.
    1980b   The Nature of Thule Eskimo Whale Use. *Arctic* 33:517-541.
    1984   History of Native Whaling in the Arctic and Subarctic. In *Arctic Whaling*, edited by H. K. s'Jacob, K. Snoeijing and R. Vaughan, pp. 79-111. Arctic Centre, University of Groningen.

McCartney, A. P., and E. D. Mitchell
    1988   Thule Eskimo Bowhead Whale Selection on Somerset Island, Arctic Canada.

Paper presented at the 53rd Annual Meeting of the Society for American Archaeology, Phoenix, April 1988.

McCartney, A. P., and J. M. Savelle
　1985　Thule Eskimo Whaling in the Central Canadian Arctic. *Arctic Anthropology* 22:37-58.

McFarland Symington, M.
　1987　Demography, Ranging Patterns and Activity Budgets of Black Spider Monkeys (*Ateles paniscus chamek*) in the Manu National Park, Peru. *American Journal of Primatology* 15:45-67.
　1988　Sex Ratio and Maternal Rank in Wild Spider Monkeys: When Daughters Disperse. *Behavioral Ecology and Sociobiology* 20:421-425.

McGhee, R.
　1969-70　Speculations on Climatic Change and Thule Culture Development. *Folk* 11/12:173-184.
　1984　The Thule Village at Brooman Point. *National Museum of Man Mercury Series, Archaeological Survey of Canada*, No. 125.

McVay, S.
　1973　Stalking the Arctic Whale. *American Scientist* 61:24-37.

Meagher, M. M.
　1973　The Bison of Yellowstone National Park. *National Park Service Scientific Monograph Series* No. 1.

Mellars, P.
　1989　Major Issues in the Emergence of Modern Humans. *Current Anthropology* 30:349-385.

Meltzer, D. J.
　1988　Late Pleistocene Human Adaptations in Eastern North America. *Journal of World Prehistory* 2:1-52.

Meltzer, D. J., and B. D. Smith
　1986　Paleoindian and Early Archaic Subsistence Strategies in Eastern North America. In *Foraging, Collecting and Harvesting: Archaic Period Subsistence and Settlement in the Eastern Woodlands*, edited by S. Neusius, pp. 1-30. Center for Archaeological Investigations, SIU, Carbondale.

Mentis, M. T.
　1972　A Review of Some Life History Features of the Large Herbivores of Africa. *The Lammergeyer* 16:1-89.

Metcalfe, D., and K. T. Jones
　1988　A Reconsideration of Animal Body-part Utility Indices. *American Antiquity* 53:486-504.

Miller, D. R.
　1975　Observations of Wolf Predation on Barren Ground Caribou in Winter. In, *Proceedings of the 1st International Reindeer and Caribou Symposium*, edited by J. R. Luick, P. C. Lent, D. R. Klein, and R. G. White, pp. 209-220. Biological Papers of the University of Alaska, Special Report No. 1, Fairbanks.

Mills, M. G. L.
　1984a　The Comparative Behavioural Ecology of the Brown Hyaena (*Hyaena brunnea*) and the Spotted Hyaena (*Crocuta crocuta*) in the Southern Kalahari. *Koedoe, Supplement* 237-247.
　1984b　Prey Selection and Feeding Habits of the Large Carnivores in the Southern Kalahari. *Koedoe, Supplement* 281-294.

Mills, M. G. L., and M. E. J. Mills
　1977　An Analysis of Bones Collected at Hyaena Breeding Dens in the Gemsbok National Parks. *Annals of the Transvaal Museum* 30:145-156.

Minor, R., K. A. Toepel, and R. L. Greenspan
　1987　Archaeological Investigations at Yaquina Head, Central Oregon Coast. *Heritage Research Associates Report*, No. 59. Eugene, Oregon.

Mitchell, B.
　1963　Determination of Age in Scottish Red Deer From Growth Layers in Dental Cementum. *Nature* 198:350-351.

Mitchell, E. D.
　1977　Initial Population Size of Bowhead Whale (*Balaena mysticetus*) Stocks: Cumulative Catch Estimates. *Report of the International Whaling Commission* 33:1-113.
　1984　Ecology of North Atlantic Boreal and Arctic Monodontid and Mysticete Whales. In *Arctic Whaling*, edited by H. K. s'Jacob, K. Snoeijing and R. Vaughan, pp. 65-78. Arctic Centre, University of Groningen.

Mitchell, E. D., and R. R. Reeves
　1981　Catch History and Cumulative Catch Estimates of Initial Population Size of Cetaceans in the Eastern Canadian Arctic. *Report of the International Whaling Commission* 31:645-682.
　1982　Factors Affecting Abundance of Bowhead Whales *Balaena mysticetus* in the Eastern Arctic of North America, 1915-1980. *Biological Conservation* 22:59-78.

Morrison, D. A.
  1983   Thule Culture in Western Coronation Gulf. *National Museum of Man Mercury Series, Archaeological Survey of Canada*, No. 116.

Murdock, G. P.
  1934   *Our Primitive Contemporaries*. Macmillan, New York.

Myers, T. P., R. G. Corner, and L. G. Tanner
  1981   Preliminary Report on the 1979 Excavations at the Clary Ranch site. *Transactions of the Nebraska Academy of Sciences* 9:1-7.

Nelson, T. A., and A. Woolf
  1987   Mortality of White-tailed Deer Fawns in Southern Illinois. *Journal of Wildlife Management* 51:326-329.

Nerini, M. K., H. W. Braham, W. M. Marquette, and D. J. Rugh
  1984   Life History of the Bowhead Whale, *Balaena mysticetus* (Mammalia: Cetacea). *Journal of Zoology* 204:443-468.

Newby, T. C.
  1973   Observations on the Breeding Behavior of the Harbor Seal in the State of Washington. *Journal of Mammalogy* 54:540-543.
  1978   Pacific Harbor Seal. In *Marine Mammals*, edited by D. Haley, pp. 185-191. Pacific Search Press, Seattle.

Nimmo, B. W.
  1971   Population Dynamics of a Wyoming Pronghorn Cohort From the Eden-Farson Site, 48SW304. *Plains Anthropologist* 16(54):285-288.

Nishimura, A., and K. Izawa
  1977   The Group Characteristics of Woolly Monkeys (*Lagothrix lagothrica*) in the Upper Amazon Basin. In *Proceedings from the Symposia of the Fifth Congress of the International Primatological Society*, edited by A. Kondo, M Kawai, A. Ehara, and S. Kawamura. Japan Science Press, Tokyo.

O'Connell, J. F., K. Hawkes, and N. Blurton Jones
  1988a Hadza Hunting, Butchering, and Bone Transport and Their Archaeological Implications. *Journal of Anthropological Research* 44:113-161.
  1988b Hadza Scavenging: Implications for Plio/Pleistocene Hominid Subsistence. *Current Anthropology* 29:356-363.

O'Connell, J. F., and B. Marshall
  1989   Analysis of Kangaroo Body Part Transport among the Alyawara of Central Australia. *Journal of Archaeological Science* 16:393-405.

Orpin, C. G.
    1982   Microbe-animal Interactions in the Rumen. In *Experimental Microbial Ecology,* edited by R. G. Burns and J. H. Slater, pp. 501-518. Blackwell Scientific Publications, Oxford.

Orr, R. T.
    1972   *Marine Mammals of California.* University of California Press, Berkeley.

Orr, R. T., and T. C. Poulter
    1967   Some Observations on Reproduction, Growth, and Social Behavior in the Steller Sea Lion. *Proceedings of the California Academy of Sciences* 35:193-226.

Payne, S.
    1973   Kill-off Patterns in Sheep and Goats: The Mandibles From Asvan Kale, *Anatolian Studies* 23:281-303.

Pearson, J. P., and B. J. Verts
    1970   Abundance and Distribution of Harbor Seals and Northern Sea Lions in Oregon. *The Murrelet* 51:1-5.

Peterson, R. O., and R. E. Page
    1988   The Rise and Fall of Isle Royale Wolves, 1975-1986. *Journal of Mammalogy* 69:89-99.

Peterson, R. O., D. L. Allen, and J. M. Dietz
    1982   Depletion of Bone Marrow Fat in Moose and a Correction for Dehydration. *Journal of Wildlife Management* 46:547-551.

Peterson, R. O., J. D. Woolington, and T. N. Bailey
    1984   Wolves of the Kenai Peninsula, Alaska. *Wildlife Monographs, No. 88.*

Peterson, R. S., and G. A Bartholomew
    1967   *The Natural History and Behavior of the California Sea Lion.* American Society of Mammalogists Special Publication No. 1.

Pianka, E. R.
    1978   *Evolutionary Ecology.* Harper and Row, New York.

Pimlott, D. H., J. A. Shannon, and G. B. Kolenosky
    1969   *The Ecology of the Timber Wolf in Algonquin Park, Ontario.* Department of Lands and Forests.

Piperno, M.
    1976-77 Analyse du Sol Moustérien de la Grotte Guattari su Mont Circé. *Quaternaria* 19:71-92.

Pitti, C., and C. Tozzi
   1971   La Grotta del Capriolo e la Buca della Iena Presso Mommio (Camaiore, Lucca). *Rivista di Scienze Preistoriche* 26:213-258.

Potts, R.
   1982   Lower Pleistocene Site Formation and Hominid Activities at Olduvai Gorge, Tanzania. Unpublished Ph.D. dissertation, Harvard University.
   1983   Foraging for Faunal Resources by Early Hominids at Olduvai Gorge, Tanzania. In *Animals and Archaeology, I. Hunters and their Prey*, edited by J. Clutton-Brock and C. Grigson, pp. 51-62. BAR International Series 163.
   1988a *Hominid Activities at Olduvai*. Aldine de Gruyter, New York.
   1988b On an Early Hominid Scavenging Niche. *Current Anthropology* 29:153-155.

Pyle, K. B.
   1980   The Cherokee Large Mammal Fauna. In *The Cherokee Excavation: Holocene Ecology and Human Adaptations in Northwestern Iowa*, edited by D. C. Anderson and H. A. Semken, Jr., pp. 171-196.

Radmilli, A. M.
   1974   *Gli Scavi nella Grotta Polesini a Ponte Lucano di Tivoli e la Più Antica Arte nel Lazio*. Sansoni Editore, Firenze.

Rapson, D. J.
   1990   Pattern and Process in Intra-site Spatial Analysis: Site Structural and Faunal Research at the Bugas-Holding Site. Unpublished Ph.D. dissertation, University of New Mexico, Albuquerque.

Rapson, D. J., and L. C. Todd
   1989   Subsistence and Structure in the Archaeological Record: Investigating Faunal Use and Distributional Patterning at the Bugas-Holding Site. Paper presented at the Symposium in Honor of Lewis R. Binford. University of New Mexico, Albuquerque, November 3-9.

Ratcliffe, P. R.
   1980   Bone Marrow Fat as an Indicator of Condition in Roe Deer. *Acta Theriologica* 25(26):333-340.

Regan, A. B.
   1934   Some Notes on the History of the Uintah Basin in Northeastern Utah to 1850. *Proceedings of the Utah Academy of Science Arts and Letters* 11:55-64.

Reher, C. A.
   1970   Appendix II: Population Dynamics of the Glenrock *Bison bison* Population. In, The Glenrock Buffalo Jump, 48CO304: Late Prehistoric Period Buffalo Procurement and Butchering on the Northwestern Plains, edited by G. C. Frison. *Plains Archaeologist Memoir* No. 7.

1973   Appendix II: The Wardell *Bison bison* Sample: Population Dynamics and Archaeological Interpretation. In, The Wardell Buffalo Trap, 48SH301: Communal Procurement in the Upper Green River Basin, Wyoming, edited by G. C. Frison. *Anthropological Papers of the Museum of Anthropology, University of Michigan* No. 48.

1974   Population Study of the Casper Site Bison. In, *The Casper Site: A Hell Gap Bison Kill on the High Plains,* edited by G. C. Frison, pp. 113-124. Academic Press, New York.

1977   Adaptive Process on the Shortgrass Plains. In *For Theory Building in Archaeology,* edited by L. R. Binford, pp. 13-40. Academic Press, New York.

Reher, C. A., and G. C. Frison
1980   The Vore Site, 48CK302, a Stratified Buffalo Jump in the Wyoming Black Hills. *Plains Anthropologist Memoir* 16.

Reeves, R. R., and S. Leatherwood
1985   Bowhead Whale: *Balaena mysticetus* Linnaeus, 1758. *Handbook of Marine Mammals, Vol. 3.* Academic Press, New York. Pp. 305-344.

Reeves, R. R., Mitchell, E. D., Mansfield, A., and M. McLaughlin
1983   Distribution and Migration of the Bowhead Whale, *Balaena mysticetus,* in the Eastern North American Arctic. *Arctic* 36:5-64.

Reich, A.
1981   Sequential Mobilization of Marrow Fat in the Impala (*Aepyceros melampus*) and Analysis of Condition of Wild Dog (*Lycaon pictus*) Prey. *Journal of Zoology (London)* 194:409-419.

Renouf, M. A. P.
1988   Sedentary Coastal Hunter-fishers: An Example From the Younger Stone Age of Northern Norway. In *The Archaeology of Prehistoric Coastlines,* edited by G. Bailey and J. Parkington, pp. 102-115. Cambridge University Press, Cambridge.

Rick, A. M.
1980   Non-cetacean Vertebrate Remains From Two Thule Winter Houses on Somerset Island, N.W.T. *Canadian Journal of Archaeology* 4:99-117.

Roberts, F. H. H.
1943   "A New Site". *American Antiquity* 8:100.

Robinson, J., and C. Janson
1987   Capuchins, Squirrel Monkeys and Atelines: Socioecological Convergence With Old World Primates. In *Primate Societies,* edited by B. Smuts, D. Cheney, R. M. Seyfarth, R. Wrangham, and T. Strusaker, pp. 69-82. University of Chicago Press, Chicago.

Rodgers, R. A., and L. D. Martin
   1984   The 12 Mile Creek Site: A Reinvestigation. *American Antiquity* 49:757-764.

Roettcher, D., and R. R. Hofmann
   1970   The Ageing of Impala from a Population in the Kenya Rift Valley. *East African Wildlife Journal* 8:37-42.

Sattenspiel, L., and H. Harpending
   1983   Stable Populations and Skeletal Age. *American Antiquity* 48:489-498.

Saunders, J. J.
   1980   A Model for Man-mammoth Relationships in Late Pleistocene North America. *Canadian Journal of Anthropology* 1:87-98.

Saunders, R. S., and J. T. Penman
   1979   Perry Ranch: A Plainview Bison Kill on the Southern Plains. *Plains Anthropologist* 24:51-65.

Savelle, J. M.
   1987   *Collectors and Foragers: Subsistence-Settlement System Change in the Central Canadian Arctic, A.D. 1000-1960*. BAR International Series 358.
   1989   Thule Eskimo Whaling in the Central Canadian Arctic: A Zooarchaeological Assessment. Report submitted to the Prince of Wales Northern Heritage Centre, Yellowknife, N.W.T.
   1990   Information Systems and Thule Eskimo Whaling. Paper presented at the Annual Meeting of the Canadian Archaeological Association, Whitehorse.

Savelle, J. M., and A. P. McCartney
   1988   Geographical and Temporal Variation in Thule Eskimo Subsistence Economies: A Model. *Research in Economic Anthropology* 10:21-72.
   1990   Prehistoric Thule Eskimo Whaling in the Central Canadian Arctic: Current Knowledge and Future Research Directions. In *Canada's Missing Dimension: Science and History in the High Arctic Islands*, edited by C. R. Harington, pp. 695-723. Canadian Museum of Nature, Ottawa.

Savishinsky, J. S., and H. S. Hara
   1981   Hare. In *Handbook of North American Indians, Volume 6: Subarctic*, edited by J. Helm, pp. 314-325. Smithsonian Institution Press, Washington, D. C.

Schaller, G. B.
   1967   *The Deer and the Tiger: A Study of Wildlife in India*. University of Chicago Press, Chicago.
   1972   *The Serengeti Lion*. University of Chicago Press, Chicago.

Schaller, G. B., and G. R. Lowther
    1969   The Relevance of Carnivore Behavior to the Study of Early Hominids. *Southwestern Journal of Anthropology* 25:307-341.

Schwarcz, H. P., W. Buhay, R. Grün, M. C. Stiner, S. Kuhn, and G. H. Miller
    n.d.   Absolute Dating of Sites in Coastal Lazio. In *Quaternaria Nuova, Supplement Volume*, edited by A. Bietti and G. Manzi. Volume in preparation.

Segre, A. G.
    1976-77   Quaternary Geology of the Palidoro Country, Rome. *Quaternaria* 19:157-161.

Semken, H. A.,Jr., and C. R. Falk
    1987   Late Pleistocene/Holocene Mammalian Faunas and Environmental Changes on the Northern Plains of the United States. In *Late Quaternary Mammalian Biogeography and Environments of the Great Plains and Prairies*, edited by R. W. Graham, H. A. Semken, Jr., and M. A. Graham, pp. 176-313. Illinois State Museum Scientific Papers, No. 22, Springfield.

Severinghaus, C. W.
    1949   Tooth Development and Wear as Criteria of Age in White-tailed Deer. *Journal of Wildlife Management* 13:195-216.

Shipman, P.
    1981   *Life History of a Fossil: An Introduction to Taphonomy and Paleoecology.* Harvard University Press, Cambridge, Mass.
    1986a   Scavenging or Hunting in Early Hominids: Theoretical Framework and Tests. *American Anthropologist* 88:27-43.
    1986b   Studies of Hominid-faunal Interactions at Olduvai Gorge. *Journal of Human Evolution* 15:691-706.

Simons, A. H., and G. Ilany
    1977   What Mean These Bones? Behavioral Implications of Gazelles' Remains from Archaeological Sites. *Paleorient* 3:269-274.

Simpson, T.
    1984   Population Dynamics of Mule Deer. *In* The Dead Indian Creek Site: An Archaic Occupation in the Absaroka Mountains of Northwestern Wyoming, edited by G. C. Frison and D. N. Walker, pp. 83-96. *Wyoming Archaeologist* 27:11-122.

Sinclair, A. R. E.
    1975   The Resource Limitation of Trophic Levels in Tropical Grassland Ecosystems. *Journal of Animal Ecology* 44:497-520.
    1977   *The African Buffalo: A Study of Resource Limitation of Populations.* University of Chicago Press, Chicago.

Sinclair, A. R. E., and P. Duncan
 1972   Indices of Condition in Tropical Ruminants. *East African Wildlife Journal* 10:143-149.

Sinclair, A. R. E., and M. Norton-Griffiths (editors)
 1979   *Serengeti: Dynamics of an Ecosystem*. University of Chicago Press, Chicago.

Skinner, J. D., S. Davis, and G. Llani
 1980   Bone Collecting by Striped Hyaenas, *Hyaena hyaena*, in Israel. *Paleont. Afr.* 23:99-104.

Smith, B. D.
 1974   Predator-prey Relationships in the Southeastern Ozarks — A. D. 1300. *Human Ecology* 2:31-43.
 1975   Middle Mississippi Exploitation of Animal Populations. *Anthropological Papers of the Museum of Anthropology*, No. 57. University of Michigan, Ann Arbor.

Smithers, R. H. N.
 1983   *The Mammals of the Southern African Subregion*. University of Pretoria, Pretoria, R.S.A.

Soffer, O.
 1985   *The Upper Paleolithic of the Central Russian Plain*. Academic Press, Orlando.

Sowls, L. K.
 1984   *The Peccaries*. University of Arizona Press, Tucson.

Speer, R. D.
 1978   Bison Remains from the Rex Rodgers Site. *In* Bison Procurement and Utilization: A Symposium, edited by L. B. Davis, and M. Wilson, pp. 113-127. *Plains Anthropologist Memoir* 14.

Speth, J. D.
 1983   *Bison Kills and Bone Counts: Decision Making by Ancient Hunters*. University of Chicago Press, Chicago.
 1987   Early Hominid Subsistence Strategies in Seasonal Habitats. *Journal of Archaeological Science* 14:13-29.
 1989   Early Hominid Hunting and Scavenging: The Role of Meat as an Energy Source. *Journal of Human Evolution* 18:329-343.
 1990   Seasonality, Resource Stress, and Food Sharing in So-called "Egalitarian" Foraging Societies. *Journal of Anthropological Archaeology* 9:148-188.

Speth, J. D., and K. A. Spielmann
  1983 Energy Source, Protein Metabolism, and Hunter-gatherer Subsistence Strategies. *Journal of Anthropological Archaeology* 2:1-31.

Spiess, A. E.
  1979 *Reindeer and Caribou Hunters: An Archaeological Study.* Academic Press, New York.

Spinage, C. A.
  1971 Geratodontology and Horn Growth of the Impala (*Aepyceros melampus*). *Journal of Zoology (London)* 164:209-225.
  1972 Age Estimation in Zebra. *East African Wildlife Journal* 10:273-277.
  1973 A Review of the Age Determination of Mammals by Means of Teeth, with Especial Reference to Africa. *East African Wildlife Journal* 11:165-187.
  1976 Age Determination of the Female Grant's Gazelle. *East African Wildlife Journal* 14:121-134.

Sponsel, L. E.
  1981 The Hunter and the Hunted in the Amazon: An Integrated Biological and Cultural Approach to the Behavioral Ecology of Human Predation. Unpublished Ph.D. dissertation, Cornell University, Ithaca, New York.

Staab, M. J.
  1979 Analysis of Faunal Material Recovered From a Thule Eskimo Site on Silumiut, N.W.T. *In* Thule Eskimo Culture: An Anthropological Retrospective, edited by A. P. McCartney, pp. 349-379. *National Museum of Man Mercury Series, Archaeological Survey of Canada*, No. 88.

Stanford, D.
  1979 Bison Kill by Ice Age Hunters. *National Geographic* 155:114-121.

Stefansson, V.
  1926 *My Life with the Eskimo.* Macmillan, New York.

Stephens, D., and J. Krebs
  1986 *Foraging Theory.* Princeton University Press, Princeton.

Steward, J.
  1938 Basin-plateau Aboriginal Sociopolitical Groups. *Bureau of American Ethnology Bulletin No. 120.*

Stiner, M. C.
  1990a The Ecology of Choice: Procurement and Transport of Animal Resources by Upper Pleistocene Hominids in West-central Italy. Unpublished Ph.D. dissertation, University of New Mexico, Albuquerque.

1990b The Use of Mortality Patterns in Archaeological Studies of Hominid Predatory Adaptations. *Journal of Anthropological Archaeology* 9:305-351.
1991a The Faunal Remains of Grotta Guattari: A Taphonomic Perspective. *Current Anthropology* 32(2):in press.
1991b Food Procurement and Transport by Human and Non-Human Predators. *Journal of Archaeological Science*:in press.

Straus, L. G.
1986 An Overview of the La Riera Chronology. In *La Riera Cave: Stone Age Hunter-gatherer Adaptations in Northern Spain*, edited by L. G. Straus and G. A. Clark, pp. 19-23. Arizona State University Anthropological Research Paper, No. 36.

Strum, S.
1983 Baboon Cues for Eating Meat. *Journal of Human Evolution* 12:327-336.

Talbot, L., and M. Talbot
1963 *The Wildebeest in Western Masailand, East Africa*. Wildlife Monographs, No. 12. The Wildlife Society.

Taschini, M.
1970 La Grotta Breuil al Monte Circeo: Per una Impostazione dello Studio del Pontiniano. *Origini* 4:45-78.

Taschini, M.
1979 L'Industrie Lithique de Grotta Guattari au Mont Circé (Latium): Definition Culturelle, Typologique et Chronologique du Pontinien. *Quaternaria* 12:179-247.

Tatum, L. S.
1980 A Seasonal Subsistence Model for Holocene Bison Hunters on the Eastern Plains of North America. In *Cherokee Excavations: Holocene Ecology and Human Adaptations in Northwestern Iowa*, edited by D. C. Anderson and H. A. Semken, Jr., pp. 149-169. Academic Press, New York.

Taylor, W. E., Jr., and R. McGhee
1979 Archaeological Material From Creswell Bay, N.W.T. *National Museum of Man Mercury Series, Archaeological Survey of Canada*, No. 85.

Teleki, G.
1981 The Omnivorous Diet and Eclectic Feeding Habits of Chimpanzees in Gombe National Park, Tanzania. In *Omnivorous Primates*, edited by R. S. O. Harding and G. Teleki, pp. 303-343. Columbia University Press, New York.

Terborgh, J.
1983 *Five New World Primates*. Princeton University Press, Princeton.

Terborgh, J., J. Fitzpatrick, and L. Emmons
  1984   Annotated Checklist of Bird and Mammal Species of Cocha Cashu Biological Station, Manu National Park, Peru. *Fieldiana Zoology* No. 21.

Thompson, D.
  1962   David Thompson's Narrative, 1784-1812. Edited by R. Glover. *Publications of the Champlain Society* No. 40. Toronto.

Thompson, I. D., and R. O. Peterson
  1988   Does Wolf Predation Alone Limit the Moose Population in Pukaskwa Park?: A Comment. *Journal of Wildlife Management* 52:556-559.

Tilson, R., F. von Blottnitz, and J. Henschel
  1980   Prey Selection by Spotted Hyaena (*Crocuta crocuta*) in the Namib Desert. *Madoqua* 12:41-49.

Todd, L. C.
  1983   The Horner Site: Taphonomy of an Early Holocene Bison Bonebed. Unpublished Ph.D. Dissertation, University of New Mexico, Albuquerque.
  1987a   Analysis of Kill-butchery Bonebeds and Interpretation of Paleoindian Hunting. In *The Evolution of Human Hunting*, edited by M. H. Nitecki and D. V. Nitecki, pp. 225-266. Plenum Press, New York.
  1987b   Taphonomy of the Horner II Bone Bed. In *The Horner Site: The Type Site of the Cody Cultural Complex*, edited by G. C. Frison and L. C. Todd, pp. 107-198. Academic Press, Orlando.

Todd, L. C., and J. L. Hofman
  1987   Bison Mandibles from the Horner and Finley Sites. In *The Horner Site: The Type Site of the Cody Cultural Complex*, edited by G. C. Frison and L. C. Todd, pp. 493-539. Academic Press, New York.

Todd, L. C., J. L. Hofman, and C. B. Schultz
  1990   Seasonality of the Scottsbluff and Lipscomb Bison Bonebeds: Implications for Modeling Paleoindian Subsistence. *American Antiquity* 55:813-827.

Todd, L. C., and D. J. Rapson
  1988a   Bonebed Analysis and Paleoindian Studies: The Mill Iron Site. Paper presented at the 53rd Annual Meeting of the Society for American Archaeology, Phoenix, Arizona, April 30.
  1988b   Long Bone Fragmentation and Interpretation of Faunal Assemblages: Approaches to Comparative Analysis. *Journal of Archaeological Science* 15: 307-325.

Todd, L. C., Witter, R. V., and G. C. Frison
  1987   Excavation and Documentation of the Princeton and Smithsonian Horner Site Assemblages. In *The Horner Site: The Type Site of the Cody Cultural*

Complex, edited by G. C. Frison and L. C. Todd, pp. 39-91. Academic Press, Orlando.

Tooby, J., and I. DeVore
1987   The Reconstruction of Hominid Behavioral Evolution Through Strategic Modelling. In *The Evolution of Human Behavior: Primate Models*, edited by W. G. Kinzey, pp. 183-237. SUNY Press, Albany.

Tozzi, C.
1970   La Grotta di S. Agostino (Gaeta). *Rivista di Scienze Preistoriche* 25:3-87.

Turner, A.
1989   Sample Selection, Schlepp Effects and Scavenging: The Implication of Partial Recovery for Interpretations of the Terrestrial Mammal Assemblage from Klasies River Mouth. *Journal of Archaeological Science* 16:1-11.

Vitagliano, S.
1984   Nota sul Pontiniano della Grotta dei Moscerini, Gaeta (Latina). *Atti della XXIV Riunione Scientifica dell'Istituto Italiano di Preistoria e Protostoria nel Lazio*, pp. 155-164. 8-11 Ottobre 1982.

Voorhies, M. R.
1969   *Taphonomy and Population Dynamics of an Early Pliocene Vertebrate Fauna, Knox County, Nebraska*. University of Wyoming Contributions to Geology Special Paper, No. 1.

Vrba, E. S.
1975   Some Evidence of Chronology and Paleoecology at Sterkfontein, Swartkrans and Kromdraai From the Fossil Bovidae. *Nature* 254:301-304.
1976   *The Fossil Bovidae of Sterkfontein, Swartkrans and Kromdraai*. Memoirs of the Transvaal Museum, No. 21.
1980   The Significance of Bovid Remains as Indicators of Environment and Predation Patterns. In *Fossils in the Making*, edited by A. K. Behrensmeyer and A. P. Hill, pp. 247-271. University of Chicago Press, Chicago.

Walker, D. N.
1986   *Studies in the Late Pleistocene Mammalian Fauna of Wyoming*. Unpublished Ph.D. dissertation, Department of Zoology and Physiology, University of Wyoming, Laramie.
1987   Late Pleistocene/Holocene Environmental Changes in Wyoming: The Mammalian Record. In *Late Quaternary Mammalian Biogeography and Environments of the Great Plains and Prairies*, edited by R. W. Graham, H. A. Semken, Jr., and M. A. Graham, pp. 334-392. Illinois State Museum Scientific Papers, No. 22, Springfield.

Wendland, W. M., A. Benn, and H. A. Semken, Jr.
1987   Evaluation of Climatic Changes on the North American Great Plains Determined From Faunal Evidence. In *Late Quaternary Mammalian Biogeography and Environments of the Great Plains and Prairies*, edited by R. W. Graham, H. A. Semken, Jr., and M. A. Graham, pp. 460-472. Illinois State Museum Scientific Papers, No. 22, Springfield.

Western, D.
1980   Linking the Ecology of Past and Present Mammal Communities, In *Fossils in the Making*, edited by A. K. Behrensmeyer and A. P. Hill, pp. 41-54. University of Chicago Press, Chicago.

Wheat, J. B.
1967   A Paleo-Indian Bison Kill. *Scientific American* 216:22-31.

White, G. C., R. A. Garrott, R. M. Bartmann, L. H. Carpenter, and A. W. Alldredge
1987   Survival of Mule Deer in Northwest Colorado. *Journal of Wildlife Management* 51:852-859.

Wilkinson, P. F.
1975   The Relevance of Musk Ox Exploitation to the Study of Prehistoric Animal Economies. In *Palaeoeconomy*, edited by E. S. Higgs, pp. 9-53. Cambridge University Press, London.
1976   "Random" Hunting and the Composition of Faunal Samples From Archaeological Excavations: A Modern Example From New Zealand. *Journal of Archaeological Science* 3:321-328.

Wilmsen, E. N., and F. H. H. Roberts
1978   *Lindenmeier, 1934-1974: Concluding Report on Investigations*. Smithsonian Contributions to Anthropology No. 24.

Wilson, G. L.
1924   The Horse and the Dog in Hidatsa Culture. *Anthropological Papers of the American Museum of Natural History* 25:127-311.

Wilson, M.
1974   The Casper Local Fauna and its Fossil Bison. In *The Casper Site: A Hell Gap Bison Kill on the High Plains*, edited by G. C. Frison, pp. 125-171. Academic Press, New York.
1980   Population Dynamics of the Garnsey Site Bison. *In* Late Prehistoric Bison Procurement in Southeastern New Mexico: The 1978 Season at the Garnsey Site (LA-18399), by J. D. Speth and W. J. Parry, pp. 88-129. *Technical Reports* No. 12; *Research Reports in Archaeology Contribution* 7, Museum of Anthropology, University of Michigan, Ann Arbor.

1983   Once Upon a River: Archaeology and Geology of the Bow River Valley at Calgary, Alberta, Canada. *National Museum of Man Mercury Series, Archaeological of Canada Paper* 114.

Wilson, M., G. C. Frison, and L. C. Todd
n.d.   Bison Dentitions from the Murray Springs Site, Arizona. Unpublished manuscript in possession of the authors.

Wolin, M. J.
1981   Fermentation in the Rumen and Human Large Intestine. *Science* 213:1463-1468.

Woolf, A., and J. D. Harder
1979   Population Dynamics of a Captive White-tailed Deer Herd With Emphasis on Reproduction and Mortality. *Wildlife Monographs, No. 67.*

Yellen, J. E.
1977   Cultural Patterning in Faunal Remains: Evidence From the !Kung Bushmen. In *Experimental Archaeology,* edited by D. Ingersoll, J. E. Yellen, W. Macdonald, pp. 271-331. Columbia University Press, New York.

Yesner, D. R.
1987   Life in the "Garden of Eden": Causes and Consequences of the Adoption of Marine Diets by Human Societies. In *Food and Evolution,* edited by M. Harris and E. B. Ross, pp. 285-310. Temple University Press, Philadelphia.
1988   Effects of Prehistoric Human Exploitation on Aleutian Sea Mammal Populations. *Arctic Anthropology* 25:28-43.

York, A. E., and J. R. Hartley
1981   Pup Production Following Harvest of Female Northern Fur Seals. *Canadian Journal of Fisheries and Aquatic Science* 38:84-90.

Yost, J., and P. Kelley
1983   Shotguns, Blowguns and Spears: The Analysis of Technological Efficiency. In *Adaptive Responses of Native Amazonians,* edited by R. Hames and W. Vickers, pp. 189-224. Academic Press, New York.

Zeimens, G. M.
1982   Analysis of the Postcranial Bison Remains. In *The Agate Basin Site,* edited by G. C. Frison and D. J. Stanford, pp. 213-240. Academic Press, New York.

List of Contributors

Michael Alvard, Department of Anthropology, University of New Mexico, Albuquerque, New Mexico 87131.

Robert J. Blumenschine, Department of Anthropology, Rutgers University, New Brunswick, New Jersey 08903.

George C. Frison, Department of Anthropology, University of Wyoming, Laramie, Wyoming 82071.

Diane Gifford-Gonzalez, Board of Studies in Anthropology, University of California at Santa Cruz, Santa Cruz, California 95064.

Jean Hudson, Department of Anthropology, University of California at Santa Barbara, Santa Barbara, California 93106.

Hillard Kaplan, Department of Anthropology, University of New Mexico, Albuquerque, New Mexico 87131.

R. Lee Lyman, Department of Anthropology, University of Missouri, Columbia, Missouri 65211.

Allen P. McCartney, Department of Anthropology, University of Arkansas, Fayetteville, Arkansas 72701.

James M. Savelle, Department of Anthropology, McGill University, 855 Sherbrooke St. West, Montreal, P.Q. H3A 2T7 Canada.

John D. Speth, Museum of Anthropology, University of Michigan, Ann Arbor, Michigan 48109.

Mary C. Stiner, Department of Anthropology, University of New Mexico, Albuquerque, New Mexico 87131.

Lawrence C. Todd, Department of Anthropology, University of New Mexico, Albuquerque, New Mexico 87131.